INTEGR

VOLUME ONE
LEBESGUE INTEGRATION AND MEASURE

LEBESGUE INTEGRATION AND MEASURE

ALAN J. WEIR

READER IN MATHEMATICS AND EDUCATION
UNIVERSITY OF SUSSEX

Published by the Press Syndicate of the University of Cambridge
The Pitt Building, Trumpington Street, Cambridge CB2 1RP
40 West 20th Street, New York, NY 10011-4211 USA
10 Stamford Road, Oakleigh, Melbourne 3166, Australia

© Cambridge University Press 1973

First published 1973
Reprinted 1985, 1987, 1988, 1991, 1994, 1996

Printed in Great Britain by Athenæum Press Ltd, Gateshead, Tyne & Wear

Library of Congress catalogue card number: 72-83584

ISBN 0 521 08728 7 hardback
ISBN 0 521 09751 7 paperback

**Transferred to
Digital Reprinting 1999**

**Printed in the
United States of America**

TO ALAN PARS

CONTENTS

	Preface	*page* ix
1	THE COMPLETENESS OF THE REALS	1
1.1	The Axiom of Completeness	2
1.2	Infima and suprema	8
1.3	Postscript on the axioms for **R**	11
2	NULL SETS	15
2.1	Countable sets	15
2.2	Null sets	18
2.3	Cantor's ternary set	20
3	THE LEBESGUE INTEGRAL ON **R**	22
3.1	Step functions	23
3.2	Construction of the Lebesgue integral	30
3.3	Relation to the 'definite integral'	44
3.4	Relation to the 'indefinite integral'	54
3.5	Some further results	63
4	THE LEBESGUE INTEGRAL ON $\mathbf{R}^k$	70
4.1	Step functions on $\mathbf{R}^k$	70
4.2	The Lebesgue integral on $\mathbf{R}^k$	77
4.3	Fubini's Theorem	83
5	THE CONVERGENCE THEOREMS	93
5.1	The Monotone Convergence Theorem	94
5.2	The Dominated Convergence Theorem	106
6	MEASURABLE FUNCTIONS AND LEBESGUE MEASURE	119
6.1	Measurable functions	119
6.2	Lebesgue measure on $\mathbf{R}^k$	124
6.3	The geometry of Lebesgue measure	134
6.4	Transformation of integrals	146

CONTENTS

7	**THE SPACES L^p**	*page* 162
7.1	The completeness of **R** as a metric space	162
7.2	The spaces L^p	164
7.3	The geometry of L^2	172
7.4	Bounded linear functionals on L^2	181
7.5	Orthonormal sets in L^2	188
7.6	Classical Fourier series	202
7.7	Reflections on Hilbert space	219
	Appendix The elements of topology	223
	Solutions	241
	References	277
	Index	279

PREFACE

The purpose of this volume is to give an introduction to the Lebesgue integral that is genuinely within the grasp of the average mathematical undergraduate: indeed the first three chapters (with the exclusion of a few short passages which are provided with an advance warning system) would be a suitable challenge to many keen students in Colleges of Education or even to some in their final year at school.

Traditionally, the exposition of this subject has followed Lebesgue and has treated *measure, measurable function* and *integral* in that order: in such an approach the student is faced with the formidable task of assimilating two subtle new concepts, each of which is attended by a fair share of technical difficulty, before he can relate the Lebesgue integral to any of the familiar integrals from his first courses in calculus. Our treatment reverses the traditional order. In the first five chapters we introduce and explore the properties of the Lebesgue integral and deduce from this, as late as Chapter 6, the consequent notions of measurable function and measure.

After a leisurely introduction to the completeness of the real line $\mathbf{R}$ in terms of increasing and decreasing sequences of real numbers, we define the integral of a step function and then approximate to the integral $\int f$ of a given function $f: \mathbf{R} \to \mathbf{R}$ by means of the integrals of increasing or decreasing sequences $\{\phi_n\}$ of step functions. This sounds exactly like the classical approach to the Riemann integral using sequences of 'lower and upper sums'; the main difference is that our sequences $\{\phi_n(x)\}$ converge to $f(x)$ for *almost all* x, i.e. there is a *null* set S such that $\phi_n(x) \to f(x)$ for every x outside S. These null sets turn out to be precisely the sets whose Lebesgue measure is zero, and it may be complained that we have not after all reversed the order of Lebesgue's original treatment. The idea of a null set is, however, much simpler than the general concept of measure, and it is treated quite briefly from first principles in Chapter 2.

The definition of $\int f$ makes no restriction on the boundedness of f and considers the values of f on the whole real line. In §3.3, when we look for simple *sufficient* conditions for the existence of $\int f$, we restrict f to the interval $[a, b]$ and assume that f is bounded. This allows us to link the Lebesgue integral with the familiar 'definite integral' or 'area under the graph of f'; the most natural sufficient condition for the

PREFACE

existence of $\int_a^b f$ in these circumstances is that the set S of discontinuities of f should be null. At this point it follows immediately from the definitions that such a function has an integral in the sense of both Riemann and Lebesgue and the two integrals are equal. As it happens, these conditions are also *necessary* for f to be Riemann-integrable on $[a, b]$ (see Ex. 3.3.F) though far from necessary for f to be Lebesgue-integrable. In §3.4 the Fundamental Theorem of the Calculus is proved in the simple case where the integrand is continuous and this facilitates the practical task of evaluating many of the standard integrals.

In Chapter 4 the Lebesgue integral on $\mathbf{R}^k$ is defined in exactly the same way and then related to integrals on $\mathbf{R}$ by means of Fubini's Theorem. This famous theorem gives an early geometrical interpretation of area and volume in terms of cross-sections which fits very well with previous experience in elementary calculus. It is noteworthy that the proof of Fubini's Theorem follows directly from the definition of the integral of f on $\mathbf{R}^k$, and does not depend on the convergence theorems which follow in Chapter 5.

We have laid particular stress on the idea of monotone sequences, and the Monotone Convergence Theorem of Chapter 5 may be regarded as the central theorem of the whole book. It guarantees a notion of completeness for the space L^1 of Lebesgue integrable functions which is closely analogous to the completeness of the real line $\mathbf{R}$. The first practical use we make of the powerful Monotone Convergence Theorem is to èvaluate the integrals of certain functions which are either unbounded or are defined on an unbounded interval. This corresponds to the study of 'improper integrals' in the Riemann theory, but here the fact that the Lebesgue integrals are defined from the beginning on the whole of $\mathbf{R}^k$, for possibly unbounded functions, simplifies the whole exercise. Lebesgue's famous Dominated Convergence Theorem is deduced from the Monotone Convergence Theorem. Among the practical corollaries of the Dominated Convergence Theorem is the justification for 'differentiating under the integral sign' which is now so straightforward that it is set as Exercise 5.2.12.

Not until Chapter 6 do we study measurable sets in detail, and even then it is by means of measurable functions. These are introduced as functions which yield integrable functions when they are suitably truncated, and the point is forcefully made that one has to look very hard to find a function that is *not* measurable. (In the second volume we shall discuss the Daniell integral and this order of presentation in

PREFACE

Chapter 6 is quite important in paving the way.) The rest of Chapter 6 is devoted to the proof of a very general statement of the Jacobian formula for the transformation of integrals (Theorem 6.4.3 or 6.4.4). To get away from the standard restrictive conditions which seem to have evolved in the context of the Riemann integral (cf. Loomis–Sternberg [16], Chapter 8) we have introduced the idea of a *density function* (defined on p. 146). This allows us to stress integration rather than differentiation: Theorem 6.4.1 is a fundamental result about integration (in which the experienced reader will recognise the density function as a Radon–Nikodym derivative). Under suitable conditions of boundedness the density function is identified as a 'measure derivative' in Theorem 6.4.2, and when the transformation is 'approximately affine' the density function is identified with the Jacobian (strictly, the absolute value of the determinant of the Jacobian matrix). We are not aware of any text written at this level which uses this approach.

In Chapter 7 we look again at the central question of completeness, this time in terms of distance rather than order. The spaces L^p ($p \geqslant 1$) which generalise L^1 are shown to be complete. This leads to some striking geometrical results, particularly for the Hilbert space L^2. These in turn shed light on the expansion of a given function in terms of orthogonal functions and on the classical theory of Fourier series. It is here that the Lebesgue integral comes into its own and its power is illustrated most vividly.

The Appendix is an introduction to the simplest ideas of general topology, illustrated for the most part by metric spaces. These ideas will play an even more important role in the second volume, but they are critical at several points in the first. The consistency of our definition of the Lebesgue integral on **R** depends on Lemma 3.2.1 which requires the Heine–Borel Theorem (Appendix, Theorem 6). In Chapter 6 when we consider the 'geometry of measure' in $\mathbf{R}^k$ we need to know two fundamental results about compact sets in $\mathbf{R}^k$ (Appendix, Theorems 7, 8). In the rather difficult proof of Proposition 6.4.1, which is vital in deriving the Jacobian transformation formula, we use a disarmingly simple result about connectedness (Appendix, Theorem 5). At several places in the text we refer to the Intermediate Value Theorem and the Mean Value Theorem, which most students will be happy to accept from earlier courses, but these results are proved for good measure in the Appendix (Theorems 3, 11) as they illustrate so nicely the ideas of connectedness and compactness.

We regard the exercises as a very important part of the book and

PREFACE

we have written out solutions or hints for all but the most straightforward of them.

The basic mathematical ideas of Chapters 3, 4 and 5 are taken from the superb book by Riesz–Nagy [21], and we have also gained much from Rudin [24] (especially his Chapter 8 on differentiation) and Asplund–Bungart [2].

Grateful thanks are due to the many patient colleagues who have discussed the material in this book and particularly to Prof. David Edmunds, Dr John Haigh, Prof. John Kingman, Prof. Walter Ledermann and Dr Alan Pars who made invaluable comments on the manuscript. It is also a pleasure to acknowledge the skill and courtesy of the Cambridge University Press in the preparation of the book.

Kind thanks also to Prof. K. L. Chung of Stanford University for pointing out that, in Exercise 5.2.2, both f_n and g_n are the derivatives of continuous functions so that the use of powerful convergence theorems is heavy-handed. He suggested the function h_n in the same exercise to rectify this; also the integral I_n in Exercise 6.1.12.

A.J.W.

1
THE COMPLETENESS OF THE REALS

We should make it clear from the beginning that we are not going to construct the real numbers from the rational numbers. On the contrary, we shall assume that there is a set **R** of real numbers in which the familiar 'algebraic' properties of addition, subtraction, multiplication, division and ordering are known to hold.† As usual, we write **Z** for the subset of **R** consisting of the *integers* $0, 1, -1, 2, -2, \ldots$ and **Q** for the set of all *rational numbers* m/n where m, n are integers and $n \neq 0$. On one small point we are unorthodox (if it is ever unorthodox to follow Bourbaki): we say that a real number x is *positive* if $x \geqslant 0$ and x is *strictly positive* if $x > 0$.

We shall also make use of the 'geometric' representation of the real numbers as points on a line. Let $|x| = x$ if $x \geqslant 0$, $|x| = -x$ if $x < 0$ and call $|x-y|$ the *distance* between the points x, y. Conventionally an *origin* is marked on the line to represent the number zero and a point distinct from the origin is marked to represent the number one. The representation is unique once these two points are given. When only one such line is under discussion it is natural to draw it across the page parallel to the lines of type and it is usual to have the point 1

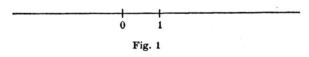

Fig. 1

on the right of the origin (Fig. 1). We refer to this line as the *real line* and denote it by the same symbol **R**. In a sense, no results are based on this representation, because one can always return to the *set* **R** and its *elements x*, but there are many advantages in having two languages for real numbers, not least the relief from monotony. The more pictorial language is a help to the intuition, especially in arguments involving the ordering of the real numbers, and it also suggests some other rather vivid terms. For example, if $a \leqslant b$, the points x satisfying $a \leqslant x \leqslant b$ form the *closed interval* $[a, b]$, the points x satisfying $a < x < b$ form the *open interval* (a, b), and the points x satisfying $a \leqslant x < b$, $a < x \leqslant b$, respectively, form the *half-open intervals* $[a, b), (a, b]$. We refer to these

† A list and brief discussion of axioms for the real numbers will be given in §1.3.

as the *bounded intervals* on **R** and give them the *length* $b-a$. We shall not follow the common practice of augmenting the real line with two 'points at infinity' $-\infty$ and ∞. Nevertheless it is often convenient to write $[a, \infty)$, (a, ∞), $(-\infty, a]$, $(-\infty, a)$ for the sets defined by $x \geq a$, $x > a$, $x \leq a$, $x < a$, respectively, and $(-\infty, \infty)$ for the whole real line **R**; we refer to these as *unbounded intervals*.

There is just one other property that we shall need, viz. the *completeness* of the real numbers. This property is essential to any thorough-going discussion of the 'analytic' ideas of limit and convergence, and these in turn are essential to any theory of integration. Our present chapter gives an elementary introduction to the completeness of **R** in terms of increasing sequences. We hope that this will help to motivate our construction of the Lebesgue integral in Chapter 3, but any reader who is already familiar with these ideas will probably be content to move on at once to Chapter 2.

1.1 The Axiom of Completeness

We shall assume a rudimentary knowledge of convergence. A *sequence* of real numbers is a function which associates a real number s_n with each integer $n \geq 1$. The number s_n, which is the *value* of the sequence for the integer n, is often called the n-th *term* of the sequence. It is usual to write the sequence as

$$s_1, s_2, s_3, \ldots$$

or simply $\{s_n\}$ for short.

In the construction of the Lebesgue integral we shall be dealing almost exclusively with *increasing* sequences and *decreasing* sequences of real numbers, i.e. those which satisfy

$$s_n \leq s_{n+1} \quad \text{for all} \quad n,$$
$$s_n \geq s_{n+1} \quad \text{for all} \quad n,$$

respectively. A sequence which is either increasing or decreasing will be called *monotone*. *Strictly increasing* and *strictly decreasing* sequences are defined with the strict inequalities.

The question of convergence for monotone sequences is quite simple, for here our geometric picture of points on the real line and our intuition make it abundantly clear what is going to happen. Suppose that

$$s_1 \leq s_2 \leq s_3 \leq \ldots,$$

then as n increases the points s_n on the real line never move to the left. They either move indefinitely far to the right, in which case we say that

1.1] AXIOM OF COMPLETENESS 3

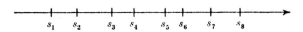

Fig. 2

$\{s_n\}$ *diverges* (to infinity), or they get nearer and nearer to some fixed point s, in which case we say that $\{s_n\}$ *converges* (to s) (see Fig. 2).

This property suggested by our intuition is exactly what we mean by the completeness of the real numbers. But we must now try to define more carefully what is meant by the last rather vague paragraph. As a first step we introduce the idea of boundedness: the sequence $\{s_n\}$ is *bounded above* if there is a real number K satisfying $s_n \leqslant K$ for all n; in this case K is called an *upper bound* of the sequence. A *divergent* increasing sequence of real numbers may now be defined as an increasing sequence which has no upper bound – more colloquially, a sequence which increases without bound. For example, the increasing sequence $1, 2, 3, \ldots$ is divergent as there is no real number K which satisfies $n \leqslant K$ for all integers n.†

In exactly the same way we say that a sequence $\{s_n\}$ of real numbers is *bounded below* if there is a real number L satisfying $s_n \geqslant L$ for all n, and we refer to L as a *lower bound* of the sequence. A decreasing sequence is *divergent* if it has no lower bound.

Let $\{s_n\}$ be a sequence of real numbers and s a real number. Recall that $\{s_n\}$ *converges* to (the *limit*) s if, given any $\epsilon > 0$, there is an integer N such that $|s_n - s| < \epsilon$ for all $n \geqslant N$.

We have now accumulated enough language to state the

Axiom of Completeness for the Real Numbers. *A sequence of real numbers which is increasing and bounded above converges to a real number.*

It is clear that a decreasing sequence $\{t_n\}$ bounded below by the real number L corresponds to an increasing sequence $\{-t_n\}$ bounded above by the real number $-L$. It therefore follows from the Axiom of Completeness that a *sequence of real numbers which is decreasing and bounded below converges to a real number.*

† This statement is often called the Axiom of Archimedes. In fact it can be deduced from our Axiom of Completeness; see Ex. 1.

A sequence like $\{1 - 1/n\}$ does not show the full power of the Axiom of Completeness because it is clear from first principles that this sequence converges to the limit 1. The main point of the Axiom is that it guarantees the *existence* of a limit for an increasing sequence provided we can prove that it is bounded above. The familiar sequence for which

$$s_n = \left(1 + \frac{1}{n}\right)^n$$

is a good illustration of this point. Let $n, r \geq 2$. The $(r+1)$-th term in the Binomial expansion of s_n is

$$\left(1 - \frac{1}{n}\right) \ldots \left(1 - \frac{r-1}{n}\right) \bigg/ r!$$

which increases strictly with n. Each term in the expansion of s_{n+1} is not less than the corresponding term in the expansion of s_n and the former expansion contains one extra positive term; thus $s_{n+1} > s_n$. The displayed term is less than $1/r!$ Thus

$$s_n < 1 + 1 + 1/2! + \ldots + 1/n! \leq 1 + 1 + \tfrac{1}{2} + \ldots + \tfrac{1}{2}^{n-1} < 3.$$

The Axiom of Completeness now assures us that $\{(1 + 1/n)^n\}$ converges to a real number between 2 and 3 (the base of natural logarithms) which we denote by e.

The above definition of convergence is not restricted to increasing or decreasing sequences of real numbers. It is convenient to gather under the heading of Proposition 1 a few of the most elementary facts about convergent sequences. As a shorthand for the statement 'the sequence $\{s_n\}$ converges to the limit s' we often write

$$s_n \to s \quad \text{or} \quad s = \lim s_n.$$

Proposition 1. *Suppose that $s_n \to s$ and $t_n \to t$. Then*
  (i) $s_n + t_n \to s + t$;
  (ii) $s_n - t_n \to s - t$;
  (iii) $s_n t_n \to st$;
  (iv) *if $t_n \neq 0$ for all n and $t \neq 0$, then $s_n/t_n \to s/t$;*
  (v) *if $s_n \geq t_n$ for all n, then $s \geq t$.*

If a sequence $\{a_n\}$ of real numbers is given, we refer to the formal expression† Σa_n as a *series* (of real numbers) whose n-th *term* is a_n.

† The reader who is unhappy about the words 'formal expression' may prefer to define a series of real numbers as a sequence $\{a_n\}$ of real numbers together with the operation of addition: thus a series is an ordered pair $(\{a_n\}, +)$ which we at once replace by the shorter notation Σa_n.

Furthermore, if the sequence $\{s_n\}$ of partial sums
$$s_n = a_1 + a_2 + \ldots + a_n$$
converges to a limit s, we say that *the series Σa_n converges to the sum s*; in keeping with historical custom and usage we shall also allow Σa_n or $a_1 + a_2 + \ldots$ to denote the sum s in this case.

It is clear that we may recapture the sequence $\{a_n\}$ from the sequence $\{s_n\}$ merely by noting that $a_1 = s_1$ and $a_n = s_n - s_{n-1}$ for $n \geq 2$. Thus the theory of sequences and the theory of series (of real numbers) are practically synonymous; it is a matter of convenience which language we adopt in a given problem.

There is one further notion of convergence which is most aptly expressed in terms of series: we say that Σa_n is *absolutely convergent* if the series $\Sigma |a_n|$ is convergent. For this terminology to make sense we must prove that an absolutely convergent series is convergent. In view of Proposition 1 the following result tells us rather more.

Proposition 2. *A series of real numbers is absolutely convergent if and only if it can be expressed as the difference of two convergent series of positive real numbers.*

Proof. (i) Let $a_n = b_n - c_n$ where b_n, c_n are positive and $\Sigma b_n, \Sigma c_n$ are convergent. Then
$$|a_1| + \ldots + |a_n| \leq b_1 + \ldots + b_n + c_1 + \ldots + c_n$$
is bounded and so $\Sigma |a_n|$ is convergent by the Axiom of Completeness.

(ii) Suppose that $\Sigma |a_n|$ is convergent. Write
$$\begin{aligned} a_n^+ &= a_n \quad \text{if } a_n \geq 0, \\ a_n^+ &= 0 \quad \text{if } a_n < 0; \\ a_n^- &= -a_n \quad \text{if } a_n \leq 0, \\ a_n^- &= 0 \quad \text{if } a_n > 0. \end{aligned}$$
Then
$$a_n = a_n^+ - a_n^-$$
and
$$|a_n| = a_n^+ + a_n^-.$$
The partial sums $|a_1| + \ldots + |a_n|$ are bounded; *a fortiori* the partial sums $a_1^+ + \ldots + a_n^+$, $a_1^- + \ldots + a_n^-$ are bounded, and so by the Axiom of Completeness the series Σa_n^+, Σa_n^- are convergent.

A familiar decimal 'expansion' such as $4 \cdot 283\ldots$ is shorthand for the sum of the series
$$4 + 2/10 + 8/10^2 + 3/10^3 + \ldots.$$

We are assured of the convergence of any such expansion because the partial sums are increasing and bounded above. In this particular case the partial sums
$$4,\ 4{\cdot}2,\ 4{\cdot}28,\ 4{\cdot}283, \ldots$$
are bounded above by 5. The Axiom of Completeness is simply stating what most of us have taken for granted every time we have written an infinite decimal expansion.

Another important fact emerges here. If an arbitrary real number k is given, we may construct a decimal expansion whose sum is k. We may as well assume that k is positive; otherwise we can find a decimal expansion for $-k$. It is helpful to refer to Fig. 3: if the points representing all the integers n are marked on the real line then the point k lies in exactly one of the half-open intervals $[n, n+1)$.† Let this interval be denoted by $[a_0, a_0+1)$ so that a_0 is the largest integer $\leqslant k$. If the interval $[a_0, a_0+1)$ is divided into ten intervals of type $[\ ,\ )$ and of equal length, then the point k lies in exactly one of these intervals. We may denote the left-hand end point of this interval by

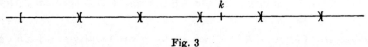

Fig. 3

$a_0 + a_1/10$, where $0 \leqslant a_1 \leqslant 9$, i.e. in decimal notation, $a_0 \cdot a_1$ is the largest (integer/10) $\leqslant k$. By further subdividing into ten half-open intervals of equal length we find $a_0 \cdot a_1 a_2$ which is the largest (integer/10^2) $\leqslant k$; and so on inductively. By our construction $a_0 \cdot a_1 a_2 \ldots a_n$ differs from k by less than $1/10^n$ and so the decimal expansion converges to k. In other words
$$k = a_0 \cdot a_1 a_2 \ldots.$$

The fact that any real number has a decimal expansion is of great practical value in all kinds of applied mathematics. It is also of profound theoretical importance because it means that one can find rational numbers (even of the type $m/10^n$) arbitrarily close to any given real number. It might be thought that one could therefore dispense with real numbers and deal with rational numbers exclusively. This would be disastrous in many ways. It was a source of anguish to the Pythagorean School in the 6th century B.C. Their idea of number in arithmetic was limited to integers and rationals; in this context they were aware that there is no (rational) number x which satisfies the simple

† We accept this as one of the elementary 'algebraic' properties of ordering on the real line; but see Ex. 5.

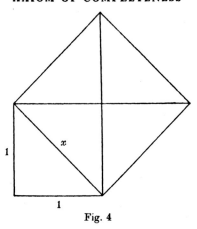

Fig. 4

equation $x^2 = 2$. In the realm of geometry, however, the most famous theorem of Pythagoras, applied to the diagonals of a unit square, provides a length x whose square is 2 (Fig. 4). The existence of such irrational or 'unutterable' numbers, as they called them, was viewed with such concern that the members were sworn to secrecy and forbidden to mention them to any outsider. Their dilemma was essentially resolved two centuries later by Eudoxus, but more than two thousand years elapsed before the mathematical world as a whole was prepared to accept the notion of the 'continuum' or, in other words, the completeness of the reals. (Using the Axiom of Completeness we can easily prove the existence of a positive *real* number whose square is 2: see Ex. 3.) In modern terms, this property of completeness is essential to the theory of limits and convergence; without it the differential calculus could not even make a beginning.

Exercises

1. Deduce the Axiom of Archimedes (footnote p. 3) from the Axiom of Completeness.

2. If $s_n \to s$ and $t_n \to t$ show that
$$\max\{s_n, t_n\} \to \max\{s, t\} \quad \text{and} \quad \min\{s_n, t_n\} \to \min\{s, t\}.$$
(Notation: if $a \geqslant b$, $\max\{a, b\} = a$, $\min\{a, b\} = b$; if $a < b$, $\max\{a, b\} = b$, $\min\{a, b\} = a$.)

3. For $n \geqslant 1$ let s_n be the largest (integer/10^{n-1}) whose square is less than 2 (e.g. $s_4 = 1 \cdot 414$). Show that $\{s_n\}$ is increasing and $\{s_n + 1/10^{n-1}\}$ is decreasing. Hence show that there is a positive real number whose square is 2.

4. If two infinite decimal expansion are different in form but converge to the same real number, show that one ends in a string of 0's and the other ends in a string of 9's (e.g. $1 \cdot 000\ldots$ and $0 \cdot 999\ldots$).

5. *The Well-Ordering Axiom* for the positive integers states that any non-empty set of positive integers has a smallest member. You are given a real number k; use the Well-Ordering Axiom and the Axiom of Archimedes (Ex. 1) to show that there is a largest integer less than or equal to k.

1.2 Infima and Suprema

Let us consider an increasing sequence $\{s_n\}$ which converges to a limit s. The diagram on the real line (Fig. 5) makes it intuitively clear that

$$s_n \leqslant s \quad \text{for all} \quad n. \tag{i}$$

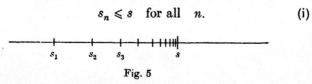

Fig. 5

We can prove this quite easily as follows. If (i) were not true, then for some N, $s_N > s$ and so all subsequent s_n, which of course satisfy $s_n \geqslant s_N$, could not get any closer to s than s_N (Fig. 6). This contradicts the definition of convergence.

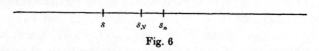

Fig. 6

Suppose now that we are given an increasing sequence $\{s_n\}$ bounded above by K. According to our Axiom of Completeness we are assured of the existence of a limit s. The diagram on the real line (Fig. 7) again makes it clear that the limit s satisfies

$$s \leqslant K. \tag{ii}$$

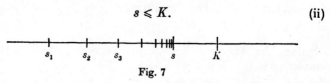

Fig. 7

The proof is quite brief. If we had $s > K$, then for s_n near enough to s we should have $s_n > K$ which would contradict the given condition $s_n \leqslant K$ for all n (see Fig. 8).

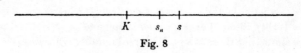

Fig. 8

Condition (i) means that s itself is an upper bound for the sequence $\{s_n\}$. On the other hand, condition (ii) tells us that no upper bound K

1.2]   INFIMA AND SUPREMA   9

can be smaller than s. We combine these two conditions quite simply by saying that s is the *least upper bound* of the sequence $\{s_n\}$. The following theorem summarises our knowledge so far of the behaviour of increasing sequences.

Theorem 1. *An increasing sequence $\{s_n\}$ of real numbers is convergent if, and only if, it is bounded above; in this case the limit of the sequence is the same as its least upper bound.*

In a much more general way, if we have an arbitrary non-empty set X of real numbers, a real number K such that $x \leqslant K$ for all x in X will be called an *upper bound* for the set X. If such a K exists we say that X is bounded above (by K). In fact K may not exist; as we have mentioned before, the set of integers $\{1, 2, 3, ...\}$ has no upper bound. If an upper bound does exist the question arises: is there a *least* upper bound? In other words, is there an upper bound B such that no number smaller than B is an upper bound? Clearly a least upper bound, assuming that it exists, is *unique*, for if there were two such least upper bounds, neither would allow the other to be smaller.

It is also clear what we mean by a *lower bound* for a non-empty set X and a *greatest lower bound*. If we write $-X$ for the set of all $-x$, where x runs through all the numbers in X, then we may use the 'reflection' $x \to -x$ which reverses the order on the real line to transfer results from X to $-X$, and vice versa. It is clear that B is the least upper bound of X if and only if $-B$ is the greatest lower bound of $-X$.

As a convenient abbreviation we shall often refer to the least upper bound of X as the *supremum* of X, the greatest lower bound of X as the *infimum* of X, and write these as sup X, inf X, for short.

Theorem 2. (i) *A non-empty set of real numbers that is bounded above has a least upper bound (supremum).*

(ii) *A non-empty set of real numbers that is bounded below has a greatest lower bound (infimum).*

Proof. In view of the above remarks it is enough to prove (ii).

Let X be a set of real numbers containing the number x and assume that X is bounded below by L, say. Of course, L may not be an integer, but there is an integer l in the interval $(L-1, L]$† which is also a lower bound for X; there are only finitely many integers in $[l, x]$† so there is a largest integer a_0 which is lower bound for X.† Among the numbers

---

† We accept these as reasonable statements about the real line; but see Ex. 3 and the footnote on p. 3.

$a_0 + r/10$ ($r = 0, 1, \ldots, 9$) there is a largest $a_0 + a_1/10$ which is a lower bound for X; in other words there is a largest (integer/10) $a_0 \cdot a_1$ which is a lower bound for X. Then there is a largest (integer/100) $a_0 \cdot a_1 a_2$ which is a lower bound for X; and so on. The expansion $a_0 \cdot a_1 a_2 \ldots$ converges to a real number s because of the Axiom of Completeness. It only remains to show that s is the greatest lower bound of X.

Consider an arbitrary element x of X. Every term of the sequence

$$a_0, \quad a_0 \cdot a_1, \quad a_0 \cdot a_1 a_2, \ldots$$

is a lower bound for X and so is less than or equal to x. We have seen above that this implies that $s \leqslant x$. Since x was arbitrary, this means that s is a lower bound for X.

Assume that $k > s$. As $1/10^n \to 0$ we may find an integer N for which $1/10^N < k - s$. In other words

$$k > s + 1/10^N$$

and so certainly $\quad k > a_0 \cdot a_1 a_2 \ldots a_N + 1/10^N$.

But this last number is too large to be a lower bound (by our construction of the sequence of decimals) and so k cannot be a lower bound. This shows finally that s is the greatest lower bound of the set X, as required.

Many mathematicians prefer to state Theorem 2 as an axiom and to deduce our Axiom of Completeness from it. (This approach is discussed very briefly in § 1.3.) Our most important reason for favouring the statement in terms of monotone sequences is that these sequences feature in our definition and main theorem for the Lebesgue integral. We also believe that many students are more confident with the idea of a bounded increasing sequence than they are with the idea of a least upper bound.

Exercises

1. If the non-empty set X of real numbers is bounded above show that $\sup X$ is an element of X if and only if X possesses a largest member.

2. Let X, Y be non-empty sets of real numbers which are bounded above. Show that
$$\sup(X \cup Y) = \max\{\sup X, \sup Y\}.$$

(For the definition of max see Ex. 1.1.2.)

Let Z be the set of all $x + y$ for x in X, y in Y. Show that
$$\sup Z = \sup X + \sup Y.$$

3. Let x be an element of X and suppose that X is bounded below by L. Use the Axiom of Archimedes (footnote on p. 3) to show that there is an integer $n_0 > x$ and an integer $n_1 < L$. Apply the Well-Ordering Axiom (Ex. 1.1.5) to the set S of all integers $n_0 - n$ for which n is a lower bound of X to find a *greatest integer lower bound* for X.

1.3 Postscript on the Axioms for R

At the beginning of this book we purposely did not indulge in a list of axioms for the real numbers. But some readers may be reassured to see how few are actually needed: it is remarkable that we only need fifteen or so (the exact number depends on the way they are formulated). Needless to say, this section is optional reading and nothing subsequently depends upon it.

First of all, to prescribe sensible laws of addition, subtraction, multiplication and division, we need the notion of a field. Suppose that F is a set containing (at least) two distinct elements 0, 1 and suppose that there are two laws of combination which associate with any two elements a, b of F a *sum* $a+b$ and a *product* ab, both in F. We shall say that F is a *field* (with respect to these operations of addition and multiplication) if the following rules are satisfied.

For any a, b, c in F,
$$a+b = b+a, \tag{1}$$
$$(a+b)+c = a+(b+c), \tag{2}$$
$$a+0 = a, \tag{3}$$

there exists an element $-a$ in F such that
$$a+(-a) = 0, \tag{4}$$
$$ab = ba, \tag{5}$$
$$(ab)c = a(bc), \tag{6}$$
$$a1 = a, \tag{7}$$

provided $a \neq 0$, there is an element a^{-1} in F such that
$$aa^{-1} = 1, \tag{8}$$
$$a(b+c) = ab+ac. \tag{9}$$

From these rules we may define the *difference* $a-b = a+(-b)$, and the *quotient* $a/b = ab^{-1}$ ($b \neq 0$), and verify all the other 'rules of algebra': for example, we may deduce that
$$(-1)(-1) = 1$$
and that $\quad (a/b)(c/d) = ac/bd \quad (b, d \neq 0)$.

We shall be content to set a few of these as exercises and leave the reader to consult an algebra text if he wishes to go into the matter in more detail.

Suppose now that F is a field and that there is a relation $\geqslant$ on F which satisfies the following conditions. For any a, b, c in F,

$$a \geqslant b \quad \text{or} \quad b \geqslant a \quad \text{(or both)}, \tag{10}$$

$$\text{if} \quad a \geqslant b \quad \text{and} \quad b \geqslant a, \quad \text{then} \quad a = b, \tag{11}$$

$$\text{if} \quad a \geqslant b \quad \text{and} \quad b \geqslant c, \quad \text{then} \quad a \geqslant c, \tag{12}$$

$$\text{if} \quad a \geqslant b, \quad \text{then} \quad a+c \geqslant b+c, \tag{13}$$

$$\text{if} \quad a \geqslant b \quad \text{and} \quad c \geqslant 0, \quad \text{then} \quad ac \geqslant bc. \tag{14}$$

Rules (10)–(12) say that $\geqslant$ is a relation of *total order* on F (regarded merely as a set) and the further rules (13), (14) say that this relation is *compatible* with addition and multiplication in the field F. In these circumstances we say that F is an *ordered field*. As an equivalent notation for $a \geqslant b$ it is convenient also to write $b \leqslant a$.

All these fourteen rules are satisfied by the rational numbers which form an ordered field **Q**. In fact, in any ordered field F the elements $0, 1, 1+1, (1+1)+1, \ldots$, which are all distinct, may be written $0, 1, 2, 3, \ldots$ and their quotients $\pm m/n$ ($n \neq 0$) give a copy of the rationals in F: in this sense **Q** is the smallest ordered field, and every ordered field contains a 'rational subfield' isomorphic to **Q** (see Ex. 10).

From now on let us assume that F is an ordered field. We need just one more axiom to ensure that F is complete. It is possible to state this in terms of increasing sequences (Ex. 11) but in the present abstract setting it is simpler to use the concept of an upper bound. A non-empty subset X of F is *bounded above* if there is an element K of F such that $x \leqslant K$ for all x in X: in this case K is an *upper bound* for X. The final axiom of completeness is:

> A non-empty subset of F that is bounded above possesses a least upper bound. (15)

In view of (11) a least upper bound of a set X, if it exists, is clearly unique: it is customary to refer to this least upper bound as the *supremum* of X or $\sup X$ for short.

By a fairly simple argument, using the idea of the rational subfield mentioned above, it is possible to prove that there cannot be more than one complete ordered field – at least, any two such are isomorphic. A more formidable task is the construction of a complete ordered field containing the field **Q** of rational numbers. This construction is given with classical beauty in G. H. Hardy's *Pure Mathematics*. We are content here to assume that such a field of real numbers exists and is unique to within isomorphism.

At a later stage (§7.1) we shall formulate the completeness of **R** in terms of distance rather than order. This allows the generalisation of the idea of completeness to metric spaces, i.e. spaces in which there is a notion of distance defined, but not necessarily any notion of order (see the first page of the Appendix).

Exercises

1. Let F_2 consist of the two elements 0, 1 and define addition and multiplication of these elements as if they were ordinary integers, save that $1+1 = 0$. Show that F_2 is a field with respect to these operations.

2. Let p be a prime number and denote by F_p the set $\{0, 1, ..., (p-1)\}$. Addition and multiplication of these elements are defined by first adding or multiplying two elements as integers and then taking the remainder when the result is divided by p. Show that F_p is a field with respect to these laws of addition and multiplication.

3. Let C consist of ordered pairs (a,b) with a,b in **R** and define
$$(a,b) + (c,d) = (a+c, b+d),$$
$$(a,b).(c,d) = (ac-bd, ad+bc).$$
Show that C is a field – the so called field of *complex numbers*. As a shorthand $(a, 0)$ is usually identified with the real number a and $(0, 1)$ is written as i; thus $i^2 = (-1, 0)$ is identified with -1 and (a,b) is written $a+ib$.

4. Show that in a field F, $a0 = 0a = 0$ for all a in F, and that
$$ab = 0 \quad \text{if and only if} \quad a = 0 \text{ or } b = 0 \quad \text{(or both)}.$$

5. If a, b are given elements of the field F show that the equation $x+b = a$ has one and only one solution, viz. $x = a-b$.

6. From the equation $(1+(-1))a = 0$ (Ex. 4) deduce that
$$(-1)a = -a.$$
Hence show that $(-1)(-1) = 1$.

7. If $a, b \neq 0$ show that $(a/b)^{-1} = b/a$. If also $d \neq 0$ show that
$$(c/d)/(a/b) = cb/da.$$

8. Let F be an ordered field and write $a > b$ if and only if $a \geqslant b$ and $a \neq b$. If $a > b$ and $b > c$ show that $a > c$.

9. Show that there is no relation of total order with respect to which F_p or C is an ordered field.

10. In the present question, as a temporary expedient, we write θ and I for the 'zero' and 'one' in the ordered field F, to distinguish them from the integers $0, 1$. Define nI for positive integers n inductively:

$$0I = \theta,$$
$$nI = (n-1)I + I \quad \text{for} \quad n \geq 1.$$

Also define $\quad (-n)I = -(nI) \quad \text{for} \quad n \geq 1.$

(i) Show that $mI = nI$ if and only if $m = n$.

(ii) Show that the set of quotients mI/nI ($n \neq 0$) form an ordered field which is a 'copy' of the ordered field $\mathbf{Q}$ of rational numbers, in the sense that the mapping
$$m/n \to mI/nI$$
preserves addition, multiplication and order.

[In this question we have tacitly assumed that certain positive integers $0, 1, \ldots$ exist and satisfy a Principle of Induction; moreover, there is a field $\mathbf{Q}$ of rational numbers which can be constructed from these positive integers. An attractive alternative is treated in [27]. For any ordered field F, we may call a subset of F *inductive* if it contains 0, and with any element x also contains $x+1$ (e.g. F is itself inductive); then the *positive integers* in F are the elements of the smallest inductive set contained in F, viz. the intersection of all inductive subsets of F. From this definition it follows at once that the Principle of Induction holds for these positive integers (and incidentally the well ordering of the positive integers may be deduced from the Principle of Induction).]

In the remaining exercises $\mathbf{R}$ *denotes a complete ordered field satisfying the rules* (1)–(15) *and the 'integers'* nI (Ex. 10) *are written simply as* n.

11. Define convergence of a sequence $\{s_n\}$ whose terms lie in $\mathbf{R}$ and deduce from (15) that an increasing sequence which is bounded above converges to its least upper bound.

12. Deduce from (15) the Archimedean property that the integers are unbounded in $\mathbf{R}$.

13. *Dedekind's Theorem.* Let $\mathbf{R}$ be expressed as the union of two non-empty disjoint subsets A, B such that $a < b$ for all a in A, b in B. Then there exists an element c of $\mathbf{R}$ such that $a \leq c \leq b$ for all a in A, b in B (c must lie in either A or B, but not in both). Deduce this result from (15).

14. *The Principle of Nested Intervals.* If $\{I_n\}$ is a sequence of closed intervals on the real line such that (i) I_n contains I_{n+1} for each $n \geq 1$, then there is at least one point contained in all the intervals I_n. If also (ii) the length of I_n tends to 0 as $n \to \infty$, then there is only one such point. Deduce this result from (15).

15. Show that the Archimedean property (Ex. 12) and the Principle of Nested Intervals (Ex. 14) together imply (15). [Adapt the proof of Theorem 1.2.2; you need the former property to show that $1/10^n \to 0$.]

2
NULL SETS

A familiar fact discussed in detail in Chapter 1 and constantly used in pure and applied mathematics is that any real number can be approximated arbitrarily closely by means of rational numbers. This may be expressed in topological terms by saying that the rational points are 'dense' on the real line. In the present chapter it is our purpose to show that the rational points are also 'sparse' on the real line! This idea of 'sparseness' is vital to an understanding of the Lebesgue integral and its correct use is one of the secrets of the power of this integral as a tool in mathematics.

2.1 Countable sets

First we must look at some questions of the 'bigness' of sets. We begin by considering the rational numbers in the interval $[0, 1]$. If we group them by means of common denominators they can be arranged as the terms of a sequence as follows:

$$0, 1, 1/2, 1/3, 2/3, 1/4, 2/4, 3/4, 1/5, 2/5, 3/5, 4/5, 1/6, \ldots.$$

The fact that $1/2$ appears again as $2/4, 3/6, 4/8, \ldots$ is not important. We can easily omit any number if it has already appeared as a term of the sequence and so obtain each rational number in $[0, 1]$ exactly once. In some ways it is less confusing to allow repeats, because the pattern of the sequence may be a little easier to recognise.

We say that a set is *countable* if its elements can be arranged as the terms of a sequence (where repeats are allowed). In this terminology, therefore, the rational numbers between 0 and 1 are countable. It is worth going a little further and proving

Theorem 1. *The rational numbers are countable.*

This can be deduced at once from the following result.

Proposition 1. *The union of a sequence of countable sets is countable.*

Proof. If the sets are $S_1, S_2, S_3, \ldots$, where

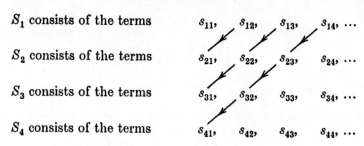

S_1 consists of the terms $s_{11}, s_{12}, s_{13}, s_{14}, \ldots$

S_2 consists of the terms $s_{21}, s_{22}, s_{23}, s_{24}, \ldots$

S_3 consists of the terms $s_{31}, s_{32}, s_{33}, s_{34}, \ldots$

S_4 consists of the terms $s_{41}, s_{42}, s_{43}, s_{44}, \ldots$

then (following the 'diagonals' as shown) the terms of the sequence

$$s_{11}, s_{12}, s_{21}, s_{13}, s_{22}, s_{31}, s_{14}, s_{23}, s_{32}, s_{41}, \ldots$$

'count' (possibly with repeats) all the elements of all the sets S_i. In other words the union $\bigcup S_i$ is countable. (Recall that the union of $S_1, S_2, \ldots$ consists of all the elements which belong to at least one of these S_i's.)

To prove Theorem 1 we have only to take for $S_1, S_2, S_3, S_4, \ldots$ the rational numbers in the intervals $[0, 1], [-1, 0], [1, 2], [-2, -1], \ldots$ respectively.

This may not strike the reader as being a particularly surprising result. In fact it may appear intuitively obvious that *all* sets are countable. The elements of any *finite* (non-empty) set can be 'counted' in the usual sense and written as $a_1, a_2, \ldots, a_N$, say. These are countable in our technical sense because they are the terms of the sequence

$$a_1, a_2, \ldots, a_N, a_1, a_2, \ldots, a_N, a_1, a_2, \ldots, a_N, \ldots$$

(and so on cyclically) or of

$$a_1, a_2, \ldots, a_N, a_N, a_N, \ldots.$$

'On the other hand,' argues the devil's advocate, 'surely no one set can be "more infinite" than another. Take the case of the reals and the rationals. Admittedly there are infinitely many real numbers between any two rationals numbers; but then, there are infinitely many rational numbers between any two real numbers! Could we not say that the score is equal and that the reals and rationals are "equally numerous"–both infinite–and leave it at that?' Mathematicians have had strong emotional reactions to these questions from the time when Cantor began to give infinite sets a much closer scrutiny. Certainly he resisted his own findings. 'This conception of the infinite is opposed to traditions which have grown dear to me, and it is much against my own will that I have been forced to accept this view. But many years of scientific speculation and trial point to the conclusions as *a logical necessity*.'†

Although we shall not need the following result it is hard to resist giving Cantor's proof (of Theorem 2) which, simple as it is, has had so many repercussions in the history of mathematics.

Theorem 2. *The real numbers are not countable.*

Proof. We shall in fact show that the real numbers in the interval $(0, 1)$ are not countable.

Let $\{s_n\}$ be an arbitrary sequence of real numbers in the open interval $(0, 1)$. The proof consists in showing that there is at least one real number in $(0, 1)$ which does *not* appear as one of these s_n's. (This is modest, to say the least, but sufficient to prove the theorem.)

These s_n's can be expressed without ambiguity as *non-terminating* decimals. (For example, we choose $0 \cdot 1999\ldots$ to represent $1/5$ rather than $0 \cdot 2$.) Let

$$s_1 = 0 \cdot a_{11} a_{12} a_{13} \ldots$$
$$s_2 = 0 \cdot a_{21} a_{22} a_{23} \ldots$$
$$s_3 = 0 \cdot a_{31} a_{32} a_{33} \ldots$$
$$\ldots$$

If $a_{nn} \neq 1$ let $b_n = 1$ and if $a_{nn} = 1$ let $b_n = 2$. This defines b_n for $n \geq 1$. Now the non-terminating decimal expansion

$$0 \cdot b_1 b_2 b_3 \ldots$$

converges to a real number b in $(0, 1)$ which differs from each s_n because its expansion differs from that of s_n in the n-th place, by our construction. (On the question of ambiguity of decimal expansions see Ex. 1.1.4.)

Exercises

1. Denote by N the set of natural numbers $\{1, 2, 3, \ldots\}$. If $S_1, S_2, S_3, \ldots$ is a sequence of subsets of N, construct a subset of N which is different from S_n for each $n \geq 1$. Hence show that the set of all subsets of N is *not* countable.

2. *Another proof of Cantor's Theorem.* Let $\{s_n\}$ be a sequence of points in the closed interval $[0, 1]$. Divide $[0, 1]$ into three equal closed intervals of length $1/3$ and let I_1 denote one of these that does not contain s_1; if in doubt choose the interval farthest to the left. Now divide I_1 into three equal closed intervals of length $1/3^2$ and let I_2 denote one of these that does not contain

† Quoted in [19] p. 208.

s_2, and so on. Use the decreasing sequence $\{I_n\}$ so constructed and the Principle of Nested Intervals (Ex. 1.3.14) to find a point x of [0, 1] distinct from every point s_n.

Why did we divide by three at each stage, rather than two?

2.2 Null sets

Even if the reader is dissatisfied with the proof of Theorem 2.1.2 and the conclusion that there are far more reals than rationals, we are now going to prove an equally graphic result which should add weight to the argument.

Select an arbitrarily small number $\epsilon > 0$. Since the rationals are countable they are the terms of a sequence $\{r_n\}$, say. Now let I_n be the open interval with *centre* r_n and *length* $l(I_n) = \epsilon/2^n$, i.e.

$$I_n = (r_n - \epsilon/2^{n+1}, r_n + \epsilon/2^{n+1}).$$

We therefore have a sequence $\{I_n\}$ of open intervals (cf. the definition of a sequence of real numbers). The *total length* of these intervals is

$$\epsilon(1/2 + 1/4 + 1/8 + \ldots) = \epsilon.$$

This means that we have covered *all* the rational numbers by means of a sequence of open intervals whose total length is arbitrarily small! It is in this sense that the rationals are 'sparse' on the real line.

Quite generally we say that a set of points on the real line is *null* if it can be covered by a sequence of (bounded) open intervals whose total length is arbitrarily small. The *total length* of the sequence $\{I_n\}$ is the sum of the series $l(I_1) + l(I_2) + \ldots$ provided this series is convergent. (No assumption is made about the intervals overlapping and no meaning is given at this stage to the 'length' of the set $\bigcup I_n$.) The set S is *covered* by $\{I_n\}$ if each point of S lies in at least one of the intervals I_n i.e. if $S \subset \bigcup I_n$.

Proposition 1. *The rational numbers form a null set.*

The proof we have given above shows in exactly the same way that

Proposition 2. *Any countable set on the real line is null.*

In practice most of the null sets we shall meet are countable, though Cantor's ternary set described in the next section is null without being countable.

This very primitive notion of null set is all that we need at present to launch into the theory of the Lebesgue integral but it is convenient to prove just one more result here.

Proposition 3. *The union of a sequence of null sets on the real line is again a null set.*

Proof. Suppose that $N_1, N_2, N_3, \ldots$ are null sets on **R**. Then, given $\epsilon > 0$; N_1 can be covered by a sequence $I_{11}, I_{12}, I_{13}, \ldots$ of open intervals of total length $\leq \epsilon/2$; N_2 can be covered by a sequence $I_{21}, I_{22}, I_{23}, \ldots$ of open intervals of total length $\leq \epsilon/2^2$; ... and so on. If we arrange these intervals in an infinite square array

$$\begin{matrix} I_{11} & I_{12} & I_{13} & \cdots \\ I_{21} & I_{22} & I_{23} & \cdots \\ I_{31} & I_{32} & I_{33} & \cdots \\ & \cdots & & \end{matrix}$$

we may 'count' down the successive diagonals (as in the proof of Proposition 2.1.1)

$$I_{11}, I_{12}, I_{21}, I_{13}, \ldots.$$

This last sequence of intervals in fact has total length $\leq \epsilon$ – we prove this in a moment – and clearly covers all the points of $N_1, N_2, \ldots$ so that the union of $N_1, N_2, \ldots$ is, by definition, null.

To prove that the total length of the sequence of intervals $I_{11}, I_{12}, I_{21}, I_{13}, \ldots$ is not greater than ϵ we note that the first n terms are certainly contained in the first n rows of the square array and so the sum of the lengths of these n terms is not greater than

$$\epsilon(1/2 + 1/4 + \ldots + 1/2^n) = \epsilon(1 - 1/2^n)$$

which in turn is less than ϵ. Theorem 1.2.1 now gives us what we want.

In our theory of integration we are going to find that null sets are in a sense 'negligible'. If a function is integrable then its values can be altered at all points of a null set without altering the value of the integral. This is one of the most practical and important properties of Lebesgue integration. Among other things it means that our theory of integration would be dealt a mortal blow if we could prove that the real numbers are countable, for then the whole real line would be null and the integral of any function equal to zero! The consistency of our theory of integration as we proceed will therefore show, very indirectly, that the real numbers are certainly not countable.

Exercise

1. Show that the word 'open' may be omitted from the definition of null sets.

2.3 Cantor's ternary set

In this brief section we shall describe a remarkable set of points on the real line. The material is purely illustrative and may be omitted at a first reading.

Consider the closed interval $[0, 1]$. If we remove the open middle third $(1/3, 2/3)$ we obtain a set S_1 which consists of two closed intervals $[0, 1/3], [2/3, 1]$ (see Fig. 9). If we now remove the open middle

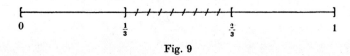

Fig. 9

thirds $(1/9, 2/9)$, $(7/9, 8/9)$ of each of the intervals of S_1 we obtain a set S_2 which consists of four closed intervals $[0, 1/9], [2/9, 1/3], [2/3, 7/9], [8/9, 1]$ (Fig. 10). Continuing in this way by removing open middle thirds we obtain a set S_n consisting of 2^n closed intervals each of length $1/3^n$.

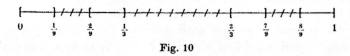

Fig. 10

Cantor's ternary set S is now defined to be the intersection of the sets S_n for $n \geqslant 1$, i.e. the set of points on the real line which lie in all the sets S_n. (As the sequence $\{S_n\}$ of sets is 'decreasing', this is a convenient way of defining the 'limit' set S.)

According to this definition S is contained in S_n which consists of (closed) intervals of total length $(2/3)^n$. As $(2/3)^n \to 0$ (and in view of Ex. 2.2.1) this means that *S is null*.

The reader will have no difficulty in adapting the material on decimal expansions in Chapter 1 to show that any real number in $[0, 1]$ has a *ternary* expansion of the form

$$\frac{a_1}{3} + \frac{a_2}{3^2} + \frac{a_3}{3^3} + \ldots,$$

where $0 \leqslant a_n \leqslant 2$ for $n \geqslant 1$. This expansion is also written

$$0 \cdot a_1 a_2 a_3 \ldots. \qquad (1)$$

Just as in Ex. 1.1.4, if two ternary expansions are different in form but converge to the same real number then one ends in a string of 0's and the other ends in a string of 2's (e.g. $0\cdot 2000\ldots = 0\cdot 1222\ldots$). This is the only kind of ambiguity that arises.

The real numbers which have a ternary expansion (1) with $a_1 = 1$ form the closed interval $[1/3, 2/3]$ (remembering that $2/3 = 0\cdot 1222\ldots$). The end points may be written as $1/3 = 0\cdot 0222\ldots$ and $2/3 = 0\cdot 2000\ldots$ and the open middle third $(1/3, 2/3)$ consists of those points for which we *must* have $a_1 = 1$. In other words S_1 consists of the points in $[0, 1]$ which have a ternary expansion with $a_1 = 0$ or 2. Exactly the same argument applies to the intervals $[0, 1/3]$, $[2/3, 1]$ and shows that S_2 consists of the points of $[0, 1]$ which have a ternary expansion with $a_1, a_2 = 0$ or 2. It is now clear, by an inductive argument, that *Cantor's ternary set consists of the real numbers in* $[0, 1]$ *which have a ternary expansion* $0\cdot a_1 a_2 a_3\ldots$ *with* $a_n = 0$ *or* 2 *for all* n.

We shall end this section by showing that S *is not countable*. The proof is reminiscent of Cantor's proof of Theorem 2.1.2.

Suppose that $s_1, s_2, s_3, \ldots$ is a sequence of numbers in Cantor's ternary set S. Then, in ternary notation,

$$s_1 = 0\cdot a_{11} a_{12} a_{13} \ldots$$

$$s_2 = 0\cdot a_{21} a_{22} a_{23} \ldots$$

$$s_3 = 0\cdot a_{31} a_{32} a_{33} \ldots$$

$$\ldots$$

where each a_{ij} is either 0 or 2. If $a_{nn} = 0$ let $b_n = 2$ and if $a_{nn} = 2$ let $b_n = 0$. Then the ternary expansion

$$0\cdot b_1 b_2 b_3 \ldots$$

converges to an element b of S and b is different from each s_n because its expansion differs from that of s_n in the n-th place, by our construction. (There is no ambiguity of expansion if we allow only 0's and 2's.)

3
THE LEBESGUE INTEGRAL ON R

A real valued function whose domain of definition is part of the real line **R** can be extended, in a simple fashion, to a function whose domain is all of **R**, by prescribing the value zero for all points not in the original domain. Any functions f to be studied in this chapter will be assumed to have domain **R** and values in **R**. In symbols

$$f: \mathbf{R} \to \mathbf{R}.$$

We shall make much of the fact that these functions form a *linear space over* **R**. In our context this simply means that any two functions f_1, f_2 and any two real numbers c_1, c_2 provide another function $c_1 f_1 + c_2 f_2$ whose value at the point x is $c_1 f_1(x) + c_2 f_2(x)$.

Most readers will have met integration in two contexts. The idea of 'indefinite integration' which gives the integral of x^2 as $x^3/3 +$ constant, will play a part later, but only confuses the issue for us at the beginning. The other idea of 'definite integration' giving the area under the graph of $y = f(x)$ is much more helpful. The integral of f which we hope to define will be a *real number* written $\int f$ (or occasionally, $\int f(x)\, dx$, when this is called for). The first property we want for $\int$ is that it should be a linear operator, i.e.

$$\int (c_1 f_1 + c_2 f_2) = c_1 \int f_1 + c_2 \int f_2 \quad \text{for} \quad c_1, c_2 \quad \text{in} \quad \mathbf{R}.$$

This equation is to be interpreted as meaning that if f_1 and f_2 are integrable then so is $c_1 f_1 + c_2 f_2$ and the two sides are equal.

Our programme is to define the integral of a very simple kind of function: one which takes the value 1 at all points of a bounded interval and zero elsewhere. The operation of integration is extended by demanding that the linear property is satisfied, and then extended once more by considering increasing and decreasing sequences. This last extension is the really subtle part and makes use of the null sets introduced in Chapter 2.

3.1 Step functions

We recall some of the most elementary notation of set theory. If a is an element of the set A we write $a \in A$, and if a is not an element of A we write $a \notin A$. If A, B are two sets and every element of A is also an element of B, then we say that A is a *subset* of B and write $A \subset B$, or equivalently, $B \supset A$.

Let $A_1, A_2, ..., A_n$ be given sets: their *union*

$$\bigcup_{i=1}^{n} A_i = A_1 \cup A_2 \cup ... \cup A_n$$

consists of all elements that belong to at least one A_i ($i = 1, 2, ..., n$), and their *intersection*

$$\bigcap_{i=1}^{n} A_i = A_1 \cap A_2 \cap ... \cap A_n$$

consists of all elements that belong to every A_i ($i = 1, 2, ..., n$).

For any two sets A, B, the *difference*

$$A \backslash B$$

consists of all elements of A that do not belong to B, and the *symmetric difference*

$$A \triangle B = (A \backslash B) \cup (B \backslash A). \ = (A \cup B) \backslash (A \cap B)$$

If S is any set of points on the real line **R** we denote by χ_S the function which takes the value 1 at all points of S and the value 0 at all other points of **R**. In other words

$$\chi_S(x) = 1 \quad \text{if} \quad x \in S,$$
$$\quad\quad = 0 \quad \text{otherwise}.$$

This function χ_S is called the *characteristic function* of S.

By a *bounded interval* on **R** we mean one of the intervals $[a, b]$, (a, b), $[a, b)$, $(a, b]$ defined (for $a \leqslant b$) at the beginning of Chapter 1; the *length* of each of these intervals is $b - a$. Provided the bounded interval I is not empty we may recapture the *end points* a, b as inf I, sup I, respectively.

The characteristic function of a bounded interval I is easy to visualise from the graph. In Figs. 11-13 a solid blob, ●, belongs to the graph and an empty ring, o, does not.

Our intuitive idea of area under the graph and formula for a rectangle 'area = base × height' suggests the following definition.

$$\int \chi_I = l(I),$$

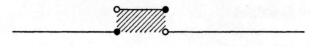

Fig. 11

where $l(I)$ stands for the length of I. The fact that this integral is the same whether I is closed, open or half-open fits with the 'obvious' fact that the area of a line segment of unit length is zero.

Our first demand of the operator $\int$ is that it should be linear. But this only makes sense if it is operating on a linear space of functions. If I_1 and I_2 are intervals and c_1, c_2 are real numbers, what kind of function is
$$\phi = c_1 \chi_{I_1} + c_2 \chi_{I_2}?$$

When the intervals I_1, I_2 are *disjoint*, i.e. their intersection $I_1 \cap I_2$ is empty, ϕ takes the constant value c_1 at all points of I_1, the constant value c_2 at all points of I_2 and the value zero elsewhere.

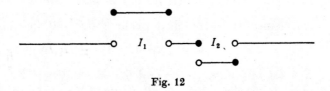

Fig. 12

When I_1, I_2 have common points their intersection $I_1 \cap I_2$ is again an interval. Furthermore, each difference
$$I_1 \backslash (I_1 \cap I_2), \quad I_2 \backslash (I_1 \cap I_2)$$
is either an interval or, at worst, the union of two disjoint intervals. The function ϕ takes the value c_1 at all points of $I_1 \backslash (I_1 \cap I_2)$, the value c_2 at all points of $I_2 \backslash (I_1 \cap I_2)$, the value $c_1 + c_2$ at all points of $I_1 \cap I_2$, and zero elsewhere.

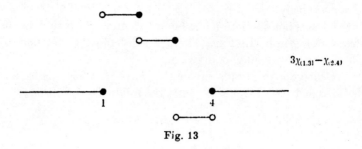

Fig. 13

3.1] STEP FUNCTIONS

Functions like $c_1\chi_{I_1} + c_2\chi_{I_2}$, given in terms of only two intervals I_1, I_2, do not yet form a linear space, because we could, for example, add two of them and get a function with even more 'steps'. But it is clear what we must do.

A *step function* is a function of the form

$$\phi = c_1\chi_{I_1} + c_2\chi_{I_2} + \ldots + c_r\chi_{I_r},$$

where $I_1, I_2, \ldots, I_r$ are bounded intervals and $c_1, c_2, \ldots, c_r$ are real numbers.

From this definition it is immediately verified that *the step functions form a linear space*:

if $\qquad \phi = c_1\chi_{I_1} + \ldots + c_r\chi_{I_r}$

and $\qquad \psi = d_1\chi_{J_1} + \ldots + d_s\chi_{J_s}$

then $\quad a\phi + b\psi = ac_1\chi_{I_1} + \ldots + ac_r\chi_{I_r} + bd_1\chi_{J_1} + \ldots + bd_s\chi_{J_s},$

which is just another linear combination of characteristic functions of bounded intervals, i.e. a step function.

In drawing the graphs of $c_1\chi_{I_1} + c_2\chi_{I_2}$, as in Figs. 12, 13, the end points of the intervals I_1, I_2 played a special role: they were the ones at which the graph had a 'jump' and the step function was constant on the intervals between these jumps. In the general case this idea yields the following useful result.

Proposition 1. *Any step function may be expressed as a (finite) linear combination of characteristic functions of disjoint intervals.*

The intervals $K_1, \ldots, K_t$ are said to be *disjoint* if every $K_i \cap K_j$ ($i \neq j$) is empty.

Proof. The step function $\phi = c_1\chi_{I_1} + \ldots + c_r\chi_{I_r}$ is given. Without loss of generality we may assume that none of the intervals I_i is empty. Suppose that they have n distinct end points; these n points divide **R** into $(2n+1)$ *disjoint* intervals, n of which are the individual points themselves, $(n-1)$ are the bounded open intervals between adjacent points, and the remaining two are unbounded intervals (see Fig. 14). As the two unbounded intervals have no points in common with

$$I_i \quad (i = 1, \ldots, r),$$

we may express I_i as the union of certain of the $(2n-1)$ bounded intervals. In this way each χ_{I_i}, and hence ϕ, is expressed as a linear combination of the characteristic functions of the $(2n-1)$ bounded intervals.

Fig. 14

As we are hoping to satisfy the linearity condition for integration, and in view of the definition of $\int \chi_I$ given above, there is now only one possible way to define the integral of a step function. If

$$\phi = c_1 \chi_{I_1} + \ldots + c_r \chi_{I_r}$$

then
$$\int \phi = c_1 l(I_1) + \ldots + c_r l(I_r).$$

This is the first of several definitions in this chapter which look obvious enough but which require careful checking as to their 'consistency'. Assuming this for a moment, the linearity is easily proved:

$$\int (a\phi + b\psi) = ac_1 l(I_1) + \ldots + ac_r l(I_r) + bd_1 l(J_1) + \ldots + bd_s l(J_s)$$

(by our definition)

$$= a(c_1 l(I_1) + \ldots + c_r l(I_r)) + b(d_1 l(J_1) + \ldots + d_s l(J_s))$$

(working in the algebra of real numbers)

$$= a \int \phi + b \int \psi$$

(by our definition).

Now for the consistency. The trouble is that a step function may be expressed in *infinitely many ways* as a linear combination of characteristic functions of intervals and it is not at once obvious that these different expressions will all give the same value for the integral. The beginner may prefer to accept the clear evidence of the 'area' under the graph of a step function and skip the rather fussy details of the rest of this section (see Ex. 10). For a deeper understanding it is important to grapple with this problem because we shall base so much upon it. The problem will occur again when we look at integration in higher dimensions where we shall deduce a *definition* and far reaching theory of area and volume.

The ambiguity can be seen clearly in two special cases:

(i) Suppose that p is a point on the bounded interval I and that I', I'', respectively, consist of the points x of I which satisfy $x < p$, $x > p$. Then
$$I = I' \cup [p,p] \cup I''$$

and so
$$\chi_I = \chi_{I'} + \chi_{[p,p]} + \chi_{I''}. \tag{1}$$

3.1]   STEP FUNCTIONS   27

The consistency of the definition of integration here amounts to the fact that
$$l(I) = l(I') + l(I'').$$

This is a fundamental property of our notion of length. If a, b are the end points of I it simply amounts to the identity
$$b - a = (p - a) + (b - p).$$

(ii) For any two real numbers c_1, c_2 and any bounded interval I,
$$(c_1 + c_2)\chi_I = c_1\chi_I + c_2\chi_I. \tag{2}$$

The question of consistency here only depends on the distributive law for real numbers which gives
$$(c_1 + c_2)l(I) = c_1 l(I) + c_2 l(I).$$

We need to show quite generally:

Proposition 2. If
$$c_1\chi_{I_1} + \ldots + c_r\chi_{I_r} = d_1\chi_{J_1} + \ldots + d_s\chi_{J_s},$$
then
$$c_1 l(I_1) + \ldots + c_r l(I_r) = d_1 l(J_1) + \ldots + d_s l(J_s). \tag{3}$$

Proof. As in the proof of Proposition 1 we assume that the intervals $I_1, \ldots, I_r$ are not empty and subdivide them into disjoint intervals using their n distinct end points. This subdivision may be carried out taking one interval and one point at a time as in case (i) above. Using equation (1) each time we subdivide and then equation (2) to gather terms involving the same interval, this gives an expression for ϕ in terms of characteristic functions of disjoint intervals:
$$\phi = e_1\chi_{K_1} + \ldots + e_t\chi_{K_t},$$
say. Since the consistency has been verified for the special cases (i) and (ii) it follows that
$$c_1 l(I_1) + \ldots + c_r l(I_r) = e_1 l(K_1) + \ldots + e_t l(K_t).$$

Let us refer to this process of introducing extra points of subdivision and the gathering of terms as 'refining' a given expression for ϕ. It is now clear that we can introduce all the end points of all the intervals $I_1, \ldots, I_r, J_1, \ldots, J_s$ and find an expression for ϕ that refines both the given ones. The two sides of (3) are therefore equal because they are both equal to the corresponding value for the common refinement. This completes the proof.

There is one property which now follows immediately from the definition of integration for step functions.

Proposition 3. *If $\phi \geq \psi$ then $\int \phi \geq \int \psi$.*

We write $f \geq g$ as an obvious shorthand for the statement
$$f(x) \geq g(x) \quad \text{for all } x \text{ in } \mathbf{R}.$$

Proof. (i) First of all let us assume that $\phi \geq 0$, i.e. $\phi(x) \geq 0$ for all x in $\mathbf{R}$. By Proposition 1 we may express ϕ as a linear combination of characteristic functions of *disjoint* intervals,
$$\phi = c_1 \chi_{I_1} + \ldots + c_r \chi_{I_r},$$
say. Since $\phi \geq 0$ it follows that each $c_i \geq 0$ and so
$$\int \phi = c_1 l(I_1) + \ldots + c_r l(I_r) \geq 0.$$

(ii) If $\phi \geq \psi$ then $\phi - \psi \geq 0$ and part (i) shows that
$$\int (\phi - \psi) \geq 0.$$
By the linearity this gives
$$\int \phi - \int \psi \geq 0,$$
i.e.
$$\int \phi \geq \int \psi$$
as required.

Before we go on to define the Lebesgue integral we should mention that Proposition 2 is not quite so trivial as it might appear. For example, if $\mathscr{R}$ denotes the collection of all sets which are finite unions of disjoint bounded intervals, we may extend the definition of length to include all the sets in $\mathscr{R}$: if
$$S = I_1 \cup \ldots \cup I_r,$$
where the intervals I_i are disjoint, then
$$l(S) = l(I_1) + \ldots + l(I_r).$$
We do not have to check the consistency of this definition from first principles as it is exactly equivalent to defining
$$l(S) = \int \chi_S,$$
where
$$\chi_S = \chi_{I_1} + \ldots + \chi_{I_r}.$$

It is easy to verify that $S \in \mathscr{R}$ if and only if χ_S is a step function (Ex. 6). With this in mind it is clear that if $S, T \in \mathscr{R}$ then

$$S \cup T, \quad S \cap T \in \mathscr{R}$$

(Ex. 6) and the (quite general) formula

$$\chi_{S \cup T} + \chi_{S \cap T} = \chi_S + \chi_T$$

yields $\quad l(S \cup T) + l(S \cap T) = l(S) + l(T).$

Another simple property of the class $\mathscr{R}$, not possessed by the class of bounded intervals on **R**, is that if $S, T \in \mathscr{R}$ then the difference $S \setminus T$ and the symmetric difference $S \triangle T$ both belong to $\mathscr{R}$ (Ex. 6). These facts are equally fundamental starting points in the rather more conventional approach to integration which goes carefully into the question of measure (in this case length) before defining any integrals.

Exercises

1. Express each of the following step functions in terms of disjoint intervals and draw their graphs.

$$\chi_{(1,4)} + 2\chi_{(2,3)}, \quad 2\chi_{(1,4)} - \chi_{(3,5)}, \quad \chi_{(1,4)} + 2\chi_{(2,3)} - 2\chi_{(3,5)}.$$

2. The function $|f|$ is defined by $|f|(x) = |f(x)|$ for all x. If ϕ is a step function show that $|\phi|$ is also a step function and that

$$\left| \int \phi \right| \leq \int |\phi|.$$

3. If ϕ, ψ are step functions show that $\max\{\phi, \psi\}$ and $\min\{\phi, \psi\}$ are step functions. (For the definition of max, min see Ex. 1.1.2 and apply this definition to the values of the functions as in the previous example.)

4. The functions f^+, f^- are defined by:

$$f^+(x) = f(x) \quad \text{if} \quad f(x) \geq 0,$$
$$= 0 \quad \text{if} \quad f(x) < 0;$$
$$f^-(x) = -f(x) \quad \text{if} \quad f(x) \leq 0,$$
$$= 0 \quad \text{if} \quad f(x) > 0.$$

Show that $f = f^+ - f^-$ and $|f| = f^+ + f^-$.
If ϕ is a step function show that ϕ^+ and ϕ^- are step functions.

5. If A, B are sets of points on **R** express the characteristic functions of $A \cup B, A \cap B, A \setminus B$ and $A \triangle B$ in terms of χ_A, χ_B (cf. Exx. 2, 3, 4).

6. Show that $S \in \mathscr{R}$ if and only if χ_S is a step function. If $S, T \in \mathscr{R}$ show that $S \cup T, S \cap T, S \setminus T, S \triangle T \in \mathscr{R}$. If $S, T \in \mathscr{R}$ and $S \subset T$ show that

$$l(S) \leq l(T).$$

7. A real number $k > 0$ and a step function ϕ are given; let S be the set of real numbers x for which $\phi(x) \geq k$. Show that $S \in \mathscr{R}$. If $\phi \geq 0$ show that

$$l(S) \leq \frac{1}{k} \int \phi.$$

8. A real number $k > 0$ and an increasing sequence $\{\phi_n\}$ of step functions are given; let S_n be the set of real numbers x for which $\phi_n(x) \geq k$. Show that $\{S_n\}$ is an increasing sequence of sets in $\mathscr{R}$.

Assume that $\phi_1 \geq 0$. If the increasing sequence $\{\int \phi_n\}$ is convergent show that the increasing sequence $\{l(S_n)\}$ is convergent.

9. The function fg is defined by: $fg(x) = f(x)g(x)$ for all x. If ϕ, ψ are step functions show that $\phi\psi$ is also a step function. By considering the step function $(a\phi + b\psi)^2$, which is positive for all real a, b, show that

$$\left(\int \phi\psi\right)^2 \leq \int \phi^2 \int \psi^2.$$

10. Assume a concept of 'area' in the plane. Define a step function ϕ in terms of its graph and give a definition of $\int \phi$ as an 'area under the graph'. Now satisfy yourself that

$$\int (a\phi + b\psi) = a\int \phi + b\int \psi$$

for any two step functions ϕ, ψ and any two real numbers a, b. (Draw the graphs of two 'smooth' positive functions f, g for which the 'area under the graph' makes sense and note that the equation

$$\int (f+g) = \int f + \int g$$

is far from obvious. See Rogosinski [**22**] p. 86 *et seq.*)

3.2 Construction of the Lebesgue integral

In order to extend our definition of integration beyond the linear space of step functions we now consider an increasing sequence $\{\phi_n\}$ of step functions, i.e. one for which $\phi_n \leq \phi_{n+1}$ for all n. In view of Proposition 3.1.3 the sequence $\{\int \phi_n\}$ is an increasing sequence of real numbers. *The only assumption we make is that this sequence of integrals is convergent.* Note that there is no assumption about the step functions ϕ_n vanishing outside some fixed interval or even that they are bounded above by some fixed constant K.

Theorem 1 (Preliminary Form). *If $\{\phi_n\}$ is an increasing sequence of step functions for which the sequence $\{\int \phi_n\}$ converges, then the points x for which the sequence $\{\phi_n(x)\}$ fails to converge form a null set.*

Proof. This is the first really searching proof in the book; but the reader who has mastered Exx. 2–8 at the end of the previous section should find little difficulty with the details.

We may assume without loss of generality that $\phi_n \geq 0$ for otherwise we could consider the increasing sequence $\{\phi_n - \phi_1\}$ and recall that $\int (\phi_n - \phi_1) = \int \phi_n - \int \phi_1$. The increasing sequence $\{\int \phi_n\}$ is convergent and so there is a real number K such that

$$\int \phi_n \leq K \quad \text{for all } n,$$

and we may obviously assume that $K > 0$.

Select an arbitrary number $\epsilon > 0$. Let S_n^ϵ consist of all the points x for which $\phi_n(x) \geq K/\epsilon$. By expressing the step function ϕ_n in terms of characteristic functions of disjoint intervals we see that S_n^ϵ may be expressed as the union of (finitely many) disjoint intervals (see Ex. 3.1.7). In terms of characteristic functions

$$(K/\epsilon)\chi_{S_n^\epsilon} \leq \phi_n$$

(using $\phi_n \geq 0$) and so $\quad (K/\epsilon) l(S_n^\epsilon) \leq \int \phi_n.$

Here we can confidently use the notion of length for S_n^ϵ introduced at the end of the previous section. But

$$\int \phi_n \leq K$$

and so $\quad (K/\epsilon) l(S_n^\epsilon) \leq K$

which implies that $\quad l(S_n^\epsilon) \leq \epsilon$
(since $K > 0$).

Now recall that $\phi_n \leq \phi_{n+1}$ and so $S_n^\epsilon \subset S_{n+1}^\epsilon$. Let S^ϵ be the union of the 'increasing' sequence of sets $\{S_n^\epsilon\}$. The first point to remark is that S^ϵ contains all the 'bad' points x for which $\{\phi_n(x)\}$ diverges. For if $\{\phi_n(x)\}$ is unbounded above, there is an integer N (depending on x and ϵ) such that $\phi_N(x) \geq K/\epsilon$. Thus x belongs to S_N^ϵ and hence to the union S^ϵ. Our theorem will be proved if we can show that S^ϵ is the union of a sequence of intervals whose total length is not greater than ϵ (see Ex. 2.2.1).

We know that $S_n^\epsilon \subset S_{n+1}^\epsilon$. The difference set $S_{n+1}^\epsilon \setminus S_n^\epsilon$ may be expressed as the union of a finite number of disjoint intervals (Ex. 3.1.6). It is now clear how we may express S^ϵ as the union of a sequence of disjoint intervals: we write down the intervals of S_1^ϵ, then those of $S_2^\epsilon \setminus S_1^\epsilon$, then those of $S_3^\epsilon \setminus S_2^\epsilon$, and so on. (In particular, the empty set qualifies as an interval.) Each point of S^ϵ will eventually occur in one of the S_n^ϵ's and so will be in one of these intervals.

The first n intervals of this sequence will certainly be contained in S_n^ϵ. But $l(S_n^\epsilon) \leq \epsilon$ for all n, and so the total length of the sequence of intervals is not greater than ϵ. This completes the proof.

There is a rather graphic language used in connection with null sets. A property $\mathfrak{P}$ which holds for all points of the real line outside some null set is said to hold *almost everywhere* (a.e.) on **R**. We shall also say that the property $\mathfrak{P}(x)$ holds for *almost all x*. The result now reads:

Theorem 1. *If $\{\phi_n\}$ is an increasing sequence of step functions for which the sequence $\{\int \phi_n\}$ converges, then $\{\phi_n\}$ converges almost everywhere.*

The last part could also read: '$\{\phi_n(x)\}$ converges for almost all x.'

Theorem 1 is absolutely fundamental in our treatment of integration. It shows how inevitably the concept of a null set arises if we begin with the situation described in the first paragraph of this section. The null sets which were introduced so abruptly in Chapter 2 now appear in a more 'natural' way because the mathematical questions we are asking about the integral force us to consider them.

We shall set aside a few more paragraphs to strengthen this conviction by proving a converse of Theorem 1 which then shows that *null sets can be characterised by the property stated in Theorem* 1. This result will not be used until the proof of Fubini's Theorem in Chapter 4 and could easily be omitted at a first reading.

Theorem 1 (Converse). *If S is a null set on **R**, then there is an increasing sequence $\{\psi_n\}$ of step functions for which the sequence $\{\int \psi_n\}$ converges and such that $\{\psi_n(x)\}$ diverges for every x in S.*

Proof. Since S is null we can find, for each integer $n \geq 1$, a sequence $I_{n1}, I_{n2}, \ldots$ of open intervals of total length $\leq 1/2^n$ which cover S. We can arrange these intervals as an infinite square array and 'count' them as in Chapter 2 by diagonals:

$$I_{11}, I_{12}, I_{21}, I_{13}, I_{22}, \ldots$$

renaming them $I_1, I_2, I_3, \ldots,$

say. Let us define
$$\psi_n = \chi_{I_1} + \ldots + \chi_{I_n}$$
for $n \geq 1$. Then, certainly $\psi_n \leq \psi_{n+1}$ for $n \geq 1$. Also
$$\int \psi_n = l(I_1) + \ldots + l(I_n) \leq 2^{-1} + 2^{-2} + \ldots + 2^{-n} < 1,$$
so that the increasing sequence $\{\int \psi_n\}$ is convergent by the Axiom of Completeness. Finally, if $x \in S$, then x is contained in some I_{ij} in the i-th row of the square array, for every $i \geq 1$. (This is just what we mean by the statement that the sequence $I_{i1}, I_{i2}, I_{i3}, \ldots$ *covers S.*) If an arbitrarily large positive integer N is given, we can go sufficiently far along the 'diagonal' sequence $\{I_n\}$ to ensure that all these I_{ij} are included for $1 \leq i \leq N$. Thus x occurs in at least N of the I_n's, and from the definition of ψ_n as a sum of characteristic functions, we see that $\psi_n(x) \geq N$ for all sufficiently large values of n. This means that $\{\psi_n(x)\}$ diverges, as required.

We now return to Theorem 1. Suppose that S is the null set consisting of the points x for which $\{\phi_n(x)\}$ fails to converge. We can define a function f_0 as follows:
$$f_0(x) = \lim \phi_n(x) \quad \text{if} \quad x \notin S,$$
$$= 0 \quad \text{if} \quad x \in S.$$

Many authors would prefer to define $f_0(x) = \infty$ when $x \in S$, because $\phi_n(x) \to \infty$ in this case, but we have chosen not to introduce at this stage a 'number' ∞; the value 0 is consistent with our convention given at the beginning of the chapter. As we shall see, f_0 can be given any values we like on the null set S without altering the definition of $\int f_0$.

Suppose that $\{\phi_n\}$ is an increasing sequence of step functions for which $\{\int \phi_n\}$ converges, and that $\phi_n(x) \to f(x)$ for almost all x, then we define the integral of f by the equation
$$\int f = \lim \int \phi_n.$$

This allows us to define the integral of the function f_0 introduced in the last paragraph, but note that f_0 is only one of infinitely many functions f to which $\{\phi_n\}$ converges almost everywhere. We return to this point in Theorem 3. As a convenient shorthand we denote by L^{inc} the set of all functions f where f is the limit almost everywhere of an increasing sequence of step functions whose integrals are bounded.

Here again there is trouble with the definition: there will be many different sequences of the kind described in Theorem 1 which converge almost everywhere to the same function f. The question of consistency this time is much harder, and any reader who is meeting these ideas for the first time would be best advised to go on to the paragraph following the proof of Lemma 2. The technical details of the next two lemmas may be less than helpful at a first reading.

The kernel of the consistency question is in the following obvious-looking result.

Lemma 1. *Let $\{\phi_n\}$ be a decreasing sequence of positive step functions converging almost everywhere to zero. Then*

$$\int \phi_n \to 0.$$

Proof. Choose $\epsilon > 0$. Suppose that ϕ_1 vanishes outside the interval $[a, b]$ and that $\phi_1 \leq K$; then the same is true for all ϕ_n. Each ϕ_n has only a finite number of points x at which there is a jump discontinuity. By Propositions 2.1.1, 2.2.2 these points (for all n) are countable and therefore form a null set A. Let B be the null set of points y for which $\{\phi_n(y)\}$ fails to converge to 0. Write C for the null set $A \cup B$ (Proposition 2.2.3) and suppose that C is covered by the sequence $\{I_n\}$ of open intervals of total length not greater than ϵ.

Consider now any point p of $[a, b]$ which is not in C. Then $\phi_n(p) \to 0$ and so there is an integer N, depending on p, for which $\phi_N(p) \leq \epsilon$. But p is not a point of discontinuity of ϕ_N and so there is an open interval J_p containing p on which ϕ_N is constant. This means that $\phi_N(x) \leq \epsilon$ for all x in J_p, and since $\{\phi_n\}$ is decreasing, $\phi_n(x) \leq \epsilon$ for all x in J_p and all $n \geq N$.

The open intervals $I_n(n \geq 1)$ and $J_p(p \notin C)$ together cover the whole interval $[a, b]$. Here we appeal to the Heine–Borel Theorem, proved in the Appendix (the proof given there may be read independently of the rest of the Appendix). This theorem shows that $[a, b]$ can be covered by a finite selection of the I_n's and J_p's: call these $I_{n_1}, ..., I_{n_r}$; $J_{p_1}, ..., J_{p_s}$. For good measure we may as well insist that there is at least one interval J_{p_1} in this finite covering and so there is a largest integer M among the numbers $N(p_1), ..., N(p_s)$. It follows that

$$\phi_n(x) \leq \epsilon \quad \text{for all } x \text{ in } \bigcup_{j=1}^{s} J_{p_j} \quad \text{and all } n \geq M.$$

3.2] LEBESGUE INTEGRAL

The last step is to show that $\int \phi_n$ is small for all sufficiently large n. Write
$$S = \bigcup_{i=1}^{r} I_{n_i} \cap [a,b],$$
$$T = \bigcup_{j=1}^{s} J_{p_j} \cap [a,b].$$

Then each of the sets S, T can be expressed as the union of a finite number of disjoint intervals; furthermore
$$l(S) \leqslant \epsilon, \quad l(T) \leqslant b-a$$
(see Ex. 3.1.6). By our construction of S and T,
$$\phi_n \leqslant K\chi_S + \epsilon\chi_T$$
for all $n \geqslant M$, and so
$$\int \phi_n \leqslant K\epsilon + \epsilon(b-a)$$
for all $n \geqslant M$. This finally gives the result.

From Lemma 1 we now deduce much more easily

Lemma 2. *Suppose that the functions f, g of L^{inc} are determined almost everywhere by increasing sequences $\{\phi_n\}$, $\{\psi_n\}$, respectively and that $f \geqslant g$ almost everywhere, then*
$$\lim \int \phi_n \geqslant \lim \int \psi_n.$$

Proof. For any fixed integer m the sequence of step functions
$$\psi_m - \phi_1, \quad \psi_m - \phi_2, \quad \psi_m - \phi_3, \ldots$$
is decreasing and converges to a negative limit almost everywhere (since $f \geqslant g$ a.e.). The sequence
$$(\psi_m - \phi_1)^+, \quad (\psi_m - \phi_2)^+, \quad (\psi_m - \phi_3)^+, \ldots$$
(Ex. 3.1.4) therefore satisfies the conditions of Lemma 1 and so
$$\int (\psi_m - \phi_n)^+ \to 0$$
as $n \to \infty$ (m fixed). But
$$\int (\psi_m - \phi_n) \leqslant \int (\psi_m - \phi_n)^+$$
by Proposition 3.1.3, and
$$\int (\psi_m - \phi_n) = \int \psi_m - \int \phi_n,$$
so
$$\int \psi_m - \lim \int \phi_n \leqslant 0.$$

Since m was arbitrary we may now let m increase and appeal to Theorem 1.2.1 to see that

$$\lim_{m\to\infty} \int \psi_m \leq \lim_{n\to\infty} \int \phi_n.$$

This is equivalent to the required result.

We can now settle the consistency: for if $\{\phi_n\}$, $\{\psi_n\}$ are two increasing sequences of step functions whose integrals are bounded above and if both sequences converge to f almost everywhere, then we may take $g = f$ in Lemma 2 and deduce

$$\lim \int \phi_n \geq \lim \int \psi_n$$

and also

$$\lim \int \phi_n \leq \lim \int \psi_n$$

from which

$$\lim \int \phi_n = \lim \int \psi_n.$$

In other words, the value of $\int f$ given by the definition does not depend on the particular sequence of step functions chosen.

The set L^{inc} of all functions which we can now integrate is in fact very large (see the last paragraph of this section), but it is not quite a linear space (Ex. 11). If $f_1, f_2 \in L^{\mathrm{inc}}$ then $c_1 f_1 + c_2 f_2 \in L^{\mathrm{inc}}$ provided c_1, c_2 are *positive* real numbers. In this case

$$\int (c_1 f_1 + c_2 f_2) = c_1 \int f_1 + c_2 \int f_2 \tag{1}$$

follows at once from the linearity of $\int$ on the step functions and Proposition 1.1.1. To find the smallest linear space containing L^{inc} we need only introduce differences: denote by L^1 the set of all functions of the form $f = g - h$ where $g, h \in L^{\mathrm{inc}}$. The reader will have no difficulty in checking that L^1 is a linear space over **R** (see the proof of Theorem 2 (i) below). We define the integral of such a function f in the obvious way by the equation

$$\int f = \int g - \int h.$$

Again there is the question of consistency, but this time, mercifully, it is easily verified. Suppose that $f = g_1 - h_1 = g_2 - h_2$ where $g_1, g_2, h_1, h_2 \in L^{\mathrm{inc}}$. Then

$$g_1 + h_2 = g_2 + h_1$$

and so, by equation (1) with $c_1 = c_2 = 1$,

$$\int g_1 + \int h_2 = \int g_2 + \int h_1$$

which gives

$$\int g_1 - \int h_1 = \int g_2 - \int h_2$$

and this is exactly what is wanted to check the consistency!

This, at last, is our definition of the *Lebesgue integral*; L^1 is called the space of (*Lebesgue*) *integrable* functions. If we wish to draw attention to the fact that we are integrating over the whole real line **R** then we may write $L^1(\mathbf{R})$ in place of L^1 (the reason for the exponent 1 will become clearer when we define the spaces L^p ($p \geqslant 1$) in Chapter 7).

We list some of the most elementary properties of the Lebesgue integral under the heading of the next theorem which we may summarise by saying that the Lebesgue integral is a *linear, positive, absolute* operator on the *linear space L^1*.

Theorem 2. (i) *If $f_1, f_2 \in L^1$ and $c_1, c_2 \in \mathbf{R}$, then $c_1 f_1 + c_2 f_2 \in L^1$ and*

$$\int (c_1 f_1 + c_2 f_2) = c_1 \int f_1 + c_2 \int f_2.$$

(ii) *If $f \in L^1$ and $f \geqslant 0$ a.e. then $\int f \geqslant 0$.*
(iii) *If $f \in L^1$ then $|f| \in L^1$ and*

$$\left| \int f \right| \leqslant \int |f|.$$

Proof. (i) Let $f_1 = g_1 - h_1$, $f_2 = g_2 - h_2$ where $g_1, g_2, h_1, h_2 \in L^{\mathrm{inc}}$ then
$$c_1 f_1 + c_2 f_2 = c_1 g_1 - c_1 h_1 + c_2 g_2 - c_2 h_2.$$

There are four cases according to the signs of c_1, c_2. We take one as typical and leave the reader to verify the others. If $c_1 \geqslant 0$, $c_2 \leqslant 0$ we write
$$c_1 f_1 + c_2 f_2 = \{c_1 g_1 + (-c_2) h_2\} - \{c_1 h_1 + (-c_2) g_2\},$$

where the terms in braces belong to L^{inc} so that $c_1 f_1 + c_2 f_2 \in L^1$ and

$$\int (c_1 f_1 + c_2 f_2) = \int \{c_1 g_1 + (-c_2) h_2\} - \int \{c_1 h_1 + (-c_2) g_2\}$$

$$= c_1 \int g_1 - c_2 \int h_2 - c_1 \int h_1 + c_2 \int g_2 \quad \text{(using (1))}$$

$$= c_1 \int f_1 + c_2 \int f_2.$$

(ii) We are given $f = g - h$ a.e. and so $g \geqslant h$ a.e. Lemma 2 gives
$$\int g \geqslant \int h,$$
i.e.
$$\int f \geqslant 0.$$

(iii) The reader is referred to Ex. 1.1.2 and also to Exx. 2, 3, 4 of §3.1. Let $f = g - h$ where $g, h \in L^{\text{inc}}$. The reader will easily verify that
$$|f| = \max\{g, h\} - \min\{g, h\}$$
(Ex. 3 below). Suppose that g, h are defined almost everywhere by increasing sequences $\{\phi_n\}$, $\{\psi_n\}$, respectively. It is convenient to arrange that ϕ_n and ψ_n are positive; this can be done, for example, by subtracting the step function $\min\{\phi_1, \psi_1\}$ from ϕ_n, ψ_n, g and h. Now $\{\max\{\phi_n, \psi_n\}\}$ and $\{\min\{\phi_n, \psi_n\}\}$ are increasing sequences of step functions which converge almost everywhere to the functions $\max\{g, h\}$ and $\min\{g, h\}$. Moreover, the integrals $\int \max\{\phi_n, \psi_n\}$, $\int \min\{\phi_n, \psi_n\}$ are dominated by $\int (\phi_n + \psi_n)$ which in turn is not greater than $\int g + \int h$. Thus $\max\{g, h\}, \min\{g, h\} \in L^{\text{inc}}$ and $|f| \in L^1$.

The last part is now easy. Since $|f| - f \geqslant 0$ we deduce from (i) and (ii) that
$$\int |f| - \int f \geqslant 0,$$
i.e.
$$\int f \leqslant \int |f|.$$

In exactly the same way
$$-\int f \leqslant \int |f|$$
and so
$$\left| \int f \right| \leqslant \int |f|.$$

One of the most useful and powerful properties of the Lebesgue integral has already been mentioned in the closing paragraph of §2.2: we can alter the values of an integrable function f at the points of a null set without altering the value of $\int f$. More precisely:

Theorem 3. *If $f_1 \in L^1$ and $f_2 = f_1$ almost everywhere, then $f_2 \in L^1$ and*
$$\int f_2 = \int f_1.$$

Proof. We have seen this already for functions of L^{inc}; it is part of the definition of $\int f$ in this case. The extension to L^1 is almost trivial.

Let $f_1 = g_1 - h_1$, where $g_1, h_1 \in L^{\mathrm{inc}}$ and define
$$g_2 = g_1, \quad h_2 = h_1 + (f_1 - f_2)$$
so that $g_2, h_2 \in L^{\mathrm{inc}}$ (the latter since $h_2 = h_1$ almost everywhere). But now
$$f_2 = g_2 - h_2$$
is in L^1 and $\int f_2 = \int g_2 - \int h_2 = \int g_1 - \int h_1 = \int f_1$ as required.

As an example, we consider the function f which takes the value 1 at the rational points of the interval $I = [0, 1]$ and vanishes everywhere else. Drawing the graph of this function is a practical impossibility, for even if we 'plot' the rational points by ink spots one millionth of an inch in diameter, they will overlap and fill out the whole interval. Thus the graph will look like Fig. 15. Nevertheless, f is a perfectly

Fig. 15

well defined function and, according to Theorem 3, has the same integral as the constant function 0 because the rational points of I form a null set. Thus $\int f = 0$.

In order to emphasise the fact that *the Lebesgue integral is on the whole of the real line* **R**, we may write:
$$\int f = \int_{\mathbf{R}} f$$
or, if you like,
$$\int f = \int_{-\infty}^{\infty} f(x)\, dx$$
in the classical notation. There is no assumption made in general about f vanishing outside a bounded interval I or even that f is bounded above or below by a constant K. It is very convenient, particularly in preparation for the next section on the 'definite integral', to introduce some further notation.

Let I be any interval (possibly unbounded) and let g be the function equal to f on I but vanishing everywhere else (see Fig. 16), i.e. $g = f\chi_I$. If $g \in L^1$ then we write
$$\int_I f = \int g$$
and refer to $\int_I f$ as the *integral of f on I*.

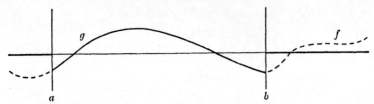

Fig. 16

The functions $f: \mathbf{R} \to \mathbf{R}$ for which $\int_I f$ can be defined in this way form a linear space, written $L^1(I)$. Thus $L^1(I)$ consists of the functions which are *integrable on I*.

In view of Theorem 3 we may ignore the end points of a bounded interval of integration. If I is any one of the four intervals $[a,b]$, (a,b), $[a,b)$, $(a,b]$ the values of $\int_I f$ will all be the same. We denote them all by

$$\int_a^b f$$

or, in the classical notation,

$$\int_a^b f(x)\,dx.$$

We have hardly ever mentioned the unbounded intervals introduced at the beginning of the first chapter. They fit very nicely into the picture here. If I is one of the unbounded intervals $[a,\infty)$, (a,∞), $\int_I f$ is written

$$\int_a^\infty f(x)\,dx,$$

and if I is one of the unbounded intervals $(-\infty, a]$, $(-\infty, a)$, $\int_I f$ is written

$$\int_{-\infty}^a f(x)\,dx.$$

There are two final results which now follow almost immediately from the definitions. From the intuitive point of view of area under a graph they are quite obvious.

Proposition 1. *Suppose that $a \leqslant c \leqslant b$. If $f \in L^1[a,c]$ and $f \in L^1[c,b]$ then $f \in L^1[a,b]$ and*

$$\int_a^b f = \int_a^c f + \int_c^b f$$

(see Fig. 17).

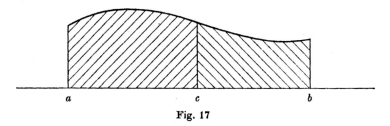

Fig. 17

Proof. This follows at once from the fact that

$$f\chi_{(a,b)} = f\chi_{(a,c)} + f\chi_{(c,b)}.$$

Proposition 2 (The Mean Value Theorem for Integrals). *Suppose that f is integrable on the interval $[a,b]$ and that there are real numbers m, M such that*

$$m \leqslant f(x) \leqslant M$$

for all x in $[a,b]$ (see Fig. 18), then

$$m(b-a) \leqslant \int_a^b f(x)\,dx \leqslant M(b-a).$$

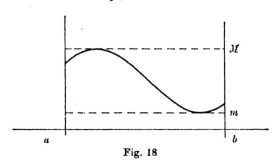

Fig. 18

Proof. We apply Theorem 2 to the inequality

$$m\chi_{(a,b)} \leqslant f\chi_{(a,b)} \leqslant M\chi_{(a,b)}.$$

We have made a comment earlier to the effect that L^{inc}, and so L^1, is 'very large'. In theory we have defined the set L^1 and should know just what functions L^1 contains, but these functions were defined rather indirectly in terms of sequences of step functions. One of our tasks in the following section is to give some recognisable (sufficient) conditions for a given function to belong to L^{inc} and so to L^1. Certainly L^{inc} contains all the functions that can be integrated in the sense of Riemann (see the remarks immediately after the statement of Theorem 3.3.1'). It also includes some unbounded functions, some functions which do not vanish outside a bounded interval and strange functions like the characteristic function of the rationals which has a discontinuity at every point. But the really significant point which will emerge as we proceed is that L^1 has a property of *completeness* which is entirely analogous to the completeness of the real numbers. The study of many branches of mathematics including the foundations of quantum mechanics, functional analysis, Fourier series and transforms, probability and statistics, to mention only a few, are as dependent upon this fact as the theory of convergence and the differential calculus are dependent upon the completeness of **R**.

Exercises

1. Evaluate the following integrals by constructing suitable increasing sequences of step functions.
$$\int_1^2 x\,dx, \quad \int_0^1 x^2\,dx.$$

2. If S is the set of irrational points between 0 and 1 find
$$\int \chi_S.$$

3. If $f = g - h$ show that
   (i) $|f| = \max\{g,h\} - \min\{g,h\}$,
   (ii) $f^+ = \max\{g,h\} - h = g - \min\{g,h\}$,
   (iii) $f^- = \max\{g,h\} - g = h - \min\{g,h\}$.

Hence show that if $f \in L^1$ then $|f|, f^+, f^- \in L^1$. (See the proof of Theorem 2 (iii).)

4. Show that $\max\{f_1, f_2\} = \tfrac{1}{2}(f_1 + f_2) + \tfrac{1}{2}|f_1 - f_2|$,
$\min\{f_1, f_2\} = \tfrac{1}{2}(f_1 + f_2) - \tfrac{1}{2}|f_1 - f_2|$.
Hence show that if $f_1, f_2 \in L^1$ then $\max\{f_1, f_2\}, \min\{f_1, f_2\} \in L^1$.

5. If $f \in L^1$ show that there is a sequence $\{\phi_n\}$ of step functions which converges to f almost everywhere and such that
$$\int |\phi_n - f| \to 0.$$

6. If $f \in L^1$ show that $f\chi_I \in L^1$ for any interval I on **R**. Deduce that $f\phi \in L^1$ for any step function ϕ.

7. If I, J are intervals on **R** and $I \subset J$ show that $L^1(I) \supset L^1(J)$. (In particular $L^1(I) \supset L^1(\mathbf{R})$.)

8. Use Ex. 7 to prove the following result (cf. Proposition 1). *Suppose that $a \leqslant c \leqslant b$. If $f \in L^1[a,b]$ then $f \in L^1[a,c]$, $f \in L^1[c,b]$ and*
$$\int_a^b f = \int_a^c f + \int_c^b f.$$

9. If $a < b$ and $f \in L^1[a,b]$ define
$$\int_b^a f = -\int_a^b f.$$
Use this convention to prove the following result (cf. Proposition 1). *Suppose that a, b, c are any three points on $\mathbb{R}$. If two of the integrals*
$$\int_a^b f, \int_a^c f, \int_c^b f$$
exist then so does the third and
$$\int_a^b f = \int_a^c f + \int_c^b f.$$

10. Let ϕ be a step function and k a non-zero real number; define
$$\psi(x) = \phi(x+k) \quad \text{for all} \quad x,$$
$$\theta(x) = \phi(kx) \quad \text{for all} \quad x.$$
Show that ψ and θ are step functions and draw their graphs in relation to the graph of ϕ.
   Show that
$$\int \psi = \int \phi,$$
$$\int \theta = \frac{1}{|k|} \int \phi.$$
If $f \in L^1$ show that $\quad \int f(x+k)\,dx = \int f(x)\,dx$

and $\quad\quad\quad\quad\quad\quad\quad \int f(kx)\,dx = \frac{1}{|k|} \int f(x)\,dx.$

What similar results can you prove if $f \in L^1[a,b]$?

11. Let $\{I_n\}$ be a sequence of open intervals in $(0,1)$ which covers all the rational points in $(0,1)$ and such that $\Sigma l(I_n) \leqslant \frac{1}{2}$ (Proposition 2.2.1). Let $S = \bigcup I_n$ and
$$f = \chi_{(0,1)} - \chi_S.$$
Show that $f \in L^1$ but that $f \notin L^{\text{inc}}$. (This proves that L^{inc} is strictly smaller than L^1, and incidentally that L^{inc} is not a linear space.)

3.3 Relation to the 'definite integral'

It is now time for us to take stock. We have fulfilled our declared programme in the introduction to this chapter and constructed an integral satisfying the simple properties listed in Theorems 3.2.2 and 3.2.3. But there may be some who are asking, 'If this is integration, what have I been doing all these years?'

As a first link with previous experience we think of the 'definite integral' as the 'area under the graph' of a function f and between the lines $x = a$, $x = b$ (Fig. 19).

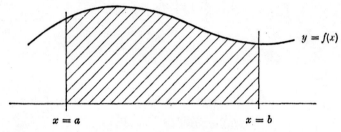

Fig. 19

For simplicity of notation we assume in the following discussion that f *vanishes outside the interval* $I = [a, b]$. For the moment we also *assume that f is positive* and approximate to f from below by a sequence $\{\phi_n\}$ of step functions obtained by successive halving of the interval I. How do we define these step functions? Suppose we have divided $[a, b)$ into two equal intervals I_1, I_2 (of type $[,)$ so that they 'fit'). We should like to take for the values of the step function ϕ_1 the smallest values of f on these two intervals. But there may not be a smallest value! The way we have chosen our intervals I_1, I_2 open on the right, there would be no smallest value even for the very smooth linear function defined by

$$f(x) = k - x \quad \text{for} \quad x \text{ in } [a, b],$$
$$= 0 \quad \text{for} \quad x \text{ outside } [a, b] \text{ (see Fig. 20)}.$$

All we need do, however, is to take

$$c_1 = \inf\{f(x) : x \in I_1\}$$
$$c_2 = \inf\{f(x) : x \in I_2\}$$

and define
$$\phi_1 = c_1 \chi_{I_1} + c_2 \chi_{I_2}.$$

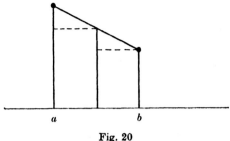

Fig. 20

These infima (greatest lower bounds) exist by the Axiom of Completeness because f is bounded below by 0.

If we now halve each of the intervals I_1, I_2 and treat them as we have treated I, we obtain the step function ϕ_2. It is clear that

$$\phi_1 \leq \phi_2$$

because the infima increase as the intervals are halved (Fig. 21).

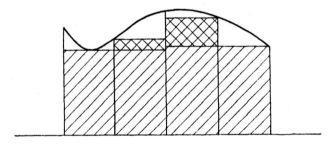

Fig. 21

By successive halving we produce an increasing sequence $\{\phi_n\}$. If we also *assume that*

$$f(x) \leq K$$

for all x, then

$$\int \phi_n \leq K(b-a)$$

for all n. (Use $\phi_n \leq K\chi_{[a,b]}$ or think of the areas as contained in a rectangular box.)

We are now in a position to quote Theorem 3.2.1 which shows that $\{\phi_n\}$ converges almost everywhere. But we want $\{\phi_n\}$ to converge almost everywhere *to our given function f*! To ensure this we make another assumption about f – that f is what we might call 'reasonably smooth'. To be precise we need the idea of *continuity*.

The function $f: \mathbf{R} \to \mathbf{R}$ is *continuous at the point p* if, given an arbitrary $\epsilon > 0$, there is a $\delta > 0$ such that

$$|f(x)-f(p)| < \epsilon \quad \text{whenever} \quad |x-p| < \delta.$$

Suppose now that f is continuous at the point p of (a,b). Given $\epsilon > 0$; there exists $\delta > 0$ such that $|f(x)-f(p)| < \epsilon$ whenever

$$|x-p| < \delta.$$

Choose N so large that $(b-a)/2^N < \delta$. One of the 2^N intervals of ϕ_N contains p and lies within the distance δ of p so that

$$f(p)-\epsilon < f(x) < f(p)+\epsilon$$

throughout this interval. By the definition of ϕ_N using infima

$$f(p)-\epsilon \leqslant \phi_N(p) \leqslant f(p).$$

But $\{\phi_n\}$ is increasing and so

$$f(p)-\epsilon \leqslant \phi_n(p) \leqslant f(p)$$

for all $n \geqslant N$. Thus we have shown that $\phi_n(p) \to f(p)$ on the assumption that p is a point of continuity of f. We want $\{\phi_n(p)\}$ to converge to $f(p)$ for almost all p and so our 'reasonably smooth' f is going to be 'continuous almost everywhere'. (Any question of the continuity of f at a or b can safely be ignored, because these two points form a null set.)

Before we state formally what we have proved it is possible to make one relaxation. We assumed that f is positive; this was to appeal to the idea of 'area' under a graph, but if we agree to count areas below the line $y = 0$ as negative, we need only *assume f is bounded*, i.e. there is a real number K such that

$$-K \leqslant f(x) \leqslant K$$

for all x in $\mathbf{R}$. This ensures the existence of infima in our definition of the sequence of step functions. It also has the advantage of treating approximations from below by increasing sequences on the same footing as approximations from above by decreasing sequences. (We are not particularly concerned about the behaviour of the sequence $\{\phi_n\}$ at the single point b but if we want to ensure that

$$\phi_n(x) \leqslant f(x) \quad \text{for all } x,$$

we may add $f(b) \chi_{[b,b]}$ to ϕ_n.) The conclusion is that both f and $-f$ belong to L^{inc} and *a fortiori* to L^1. The formal result now reads:

Theorem 1. *Let f be a function which vanishes outside the interval $[a, b]$. If f is bounded and if the points of discontinuity of f form a null set, then $f \in L^1$.*

The condition that f should vanish outside $[a, b]$ may seem artificial in the setting of the 'definite integral'. It is dictated by our convention in the opening sentences of this chapter to define f on the whole of **R**, and merely amounts to the fact that the values of f outside $[a, b]$ are ignored in the process of 'integrating from a to b'. If we use the notation $L^1[a, b]$ introduced at the end of the previous section, this result can be given a more familiar look.

Theorem 1'. *If f is bounded on $[a, b]$ and if the points of discontinuity of f on $[a, b]$ form a null set, then $f \in L^1[a, b]$.*

It is hoped that the reader will now agree that our definition of integration fits the intuitive idea of the area under a graph where this area can be found by dividing it into narrower and narrower vertical strips and approximating from below (or above).

The reader who has had an introductory course in Riemann integration will recognise the integrals $\int \phi_n$ of the increasing sequence of step functions $\{\phi_n\}$ as 'lower sums' which approximate to the 'integral of f' from below. We have seen that the increasing sequence $\{\int \phi_n\}$ converges to $\int f$ (in the sense of Lebesgue). If we apply exactly the same argument to $-f$ we find a decreasing sequence $\{\psi_n\}$ of step functions for which $\int \psi_n \to \int f$. Taking these results together we see that *a function which satisfies the conditions of Theorem 1' is integrable in the sense of Riemann and both theories give the same value for the integral of f from a to b*. In fact, the conditions of Theorem 1' are also *necessary* for the integrability of f in the sense of Riemann (this does not apply to the 'improper' Riemann integrals: see the end of §5.1). We shall make no further reference to Riemann integration in the main text but anyone who is interested in proving this result and in linking the definitions of Riemann and Lebesgue is invited to look at Exercises A to F at the end of this section which are very difficult as exercises but have sketch solutions at the end of the book. (The results extend almost without change to the higher dimensional spaces $\mathbf{R}^k$.)

There are two very important special cases of Theorem 1' that we must now mention as they are the ones which most commonly arise in practice.

We shall say that the function f is *continuous on the interval I* if the restriction of f to the interval I is continuous. More specifically this means that for any point p of I and any $\epsilon > 0$, there exists a $\delta > 0$ such that $|f(x) - f(p)| < \epsilon$ whenever $|x - p| < \delta$ and $x \in I$.

If I is the open interval (a, b) this last proviso can be ignored because δ can always be chosen, once p is given, so that the whole interval $(p - \delta, p + \delta)$ lies in I. But if I is the closed interval $[a, b]$ we have a restricted kind of continuity at the points a, b: *continuity from the right* and *continuity from the left*, respectively (Fig. 22).

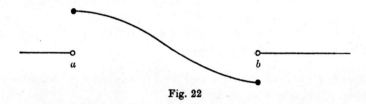

Fig. 22

We prove in the Appendix that a function which is continuous on a bounded *closed* interval $[a, b]$ is bounded on $[a, b]$. The following result is therefore a special case of Theorem 1'.

Corollary 1. *If f is continuous on $[a, b]$ then f is integrable on $[a, b]$.*

A function f is said to be *increasing* if $x \leq y$ implies $f(x) \leq f(y)$. More generally, f is *increasing on the interval I* if $x \leq y$ and $x, y \in I$ imply $f(x) \leq f(y)$. The definition of a decreasing function is entirely similar. A function which is either increasing or decreasing is called *monotone*.

One of the most easily visualised properties of a monotone function is that its discontinuities, if any, are simple 'jumps' (Fig. 23).

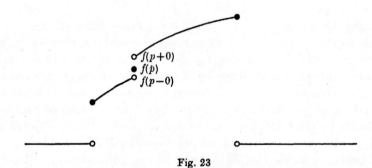

Fig. 23

Let f be an increasing function and consider the point p. Certainly $f(x) \leq f(p)$ for $x < p$ and we may write

$$f(p - 0) = \sup\{f(x) : x < p\}.$$

It follows that $f(p-0) \leq f(p)$.

Similarly $f(p+0) = \inf\{f(x): x > p\}$.

Thus $f(p-0) \leq f(p) \leq f(p+0)$. (1)

When f is decreasing we give similar definitions:
$$f(p-0) = \inf\{f(x): x < p\},$$
$$f(p+0) = \sup\{f(x): x > p\}.$$

Proposition 1. *The monotone function f is continuous at the point p if, and only if, $f(p-0) = f(p+0)$.*

Proof. We assume that f is increasing; when f is decreasing the proof is entirely similar.

(i) Suppose that f is continuous at the point p. Then there are values of $f(x)$ arbitrarily close to $f(p)$ on either side and so
$$\sup\{f(x): x < p\} = \inf\{f(x): x > p\} = f(p).$$

(ii) Suppose that $f(p-0) = f(p+0)$. Then each equals $f(p)$ by (1). Given $\epsilon > 0$; $f(p) - \epsilon < f(p)$ and so $f(p) - \epsilon$ is too small to be an upper bound for the values $f(x)$ ($x < p$). Thus there is a point $q < p$ for which $f(p) - \epsilon < f(q)$. Similarly, there is a point $r > p$ for which $f(r) < f(p) + \epsilon$. Using the fact that f is increasing,
$$f(p) - \epsilon < f(x) < f(p) + \epsilon$$
for all x in the interval $[q, r]$. We can take δ equal to $\min\{p-q, r-p\}$ and get
$$f(p) - \epsilon < f(x) < f(p) + \epsilon$$
for $|x - p| < \delta$. In other words, f is continuous at the point p.

According to this elementary proposition any discontinuity p of a monotone function f involves a vertical jump described by a vertical interval with end points $f(p-0), f(p+0)$. All we need here is to note that each of these intervals certainly contains a rational point distinct from its end points and these rational points can be used to 'count' the discontinuities. In view of Proposition 2.2.2 this means that the discontinuities form a null set and we can use Theorem 1' again.

Corollary 2. *If f is monotone on $[a, b]$ then f is integrable on $[a, b]$.*

All this is very satisfactory. We have found some practical *sufficient* conditions in Theorem 1 and Corollaries for functions to be integrable. It is also worth repeating the remarks made at the end of the last section that the Lebesgue integral, as we have defined it, is not restricted by *any* of the conditions of Theorem 1. The characteristic function of the rationals which is equal to zero almost everywhere is certainly integrable but is discontinuous at every point of the real line. Nor is there any reason to suppose that a function f of L^1 is bounded either above or below by a constant K. We shall see plenty of examples of this in the sequel but we can construct an artificial one by defining $f(x) = x$ for integral values of x and $f(x) = 0$ for all other values of x.

We now have a considerable repertoire of functions which are known to be integrable. But how are we to *evaluate* their integrals? It is to this practical problem that the next section gives a partial answer as it links the ideas of integration and differentiation.

Exercises

1. If f and g are continuous at the point p show that $f+g$, $f-g$, fg are continuous at p; also f/g is continuous at p provided $g(p) \neq 0$.

2. Use Ex. 1 to prove that the *polynomial function* f defined by

$$f(x) = a_0 x^n + a_1 x^{n-1} + \ldots + a_n$$

for all x in $\mathbf{R}$ (all $a_i \in \mathbf{R}$) is continuous on $\mathbf{R}$.
If
$$g(x) = b_0 x^m + b_1 x^{m-1} + \ldots + b_m$$

(all $b_j \in \mathbf{R}$, $b_0 \neq 0$) show that the *rational function* h defined by

$$h(x) = f(x)/g(x) \quad \text{for} \quad g(x) \neq 0,$$
$$= 0 \quad \text{for} \quad g(x) = 0$$

is continuous at any point p for which $g(p) \neq 0$.

3. Show that $\int_1^x \frac{1}{t} dt$ exists for $x > 0$.

(See Ex. 3.2.9 for the definition of this integral when $0 < x < 1$.)

Prove that $\quad \dfrac{x-1}{x} \leqslant \int_1^x \frac{1}{t} dt \leqslant x-1 \quad \text{for} \quad x > 0.$

4. If $n \geqslant 2$ is an integer show that

$$\frac{1}{2} + \frac{1}{3} + \ldots + \frac{1}{n} < \int_1^n \frac{1}{t} dt < 1 + \frac{1}{2} + \ldots + \frac{1}{n-1}.$$

5. Given two functions f, g (of **R** into **R**) the *composite function* $h = f \circ g$ is defined by the equation
$$h(x) = f(g(x))$$
for all x in **R**. Show that if g is continuous at the point p and if f is continuous at the point $g(p)$, then h is continuous at the point p.

Exercises on the Riemann Integral

The purpose of these exercises is to give a brief outline of the definitions and basic results for the Riemann integral. The first four are part of 'standard' Riemann theory and use only the elementary integrals of step functions defined in §3.1. The remaining two exercises link the Riemann and Lebesgue integrals; they use the theory developed in the present section and anticipate one fundamental fact from Chapter 5: *if a positive function has Lebesgue integral equal to zero then the function is zero almost everywhere*.

Throughout these exercises $f: \mathbf{R} \to \mathbf{R}$ is a *bounded* function which vanishes outside the (bounded) interval $[a, b]$. The former is an essential condition; the latter is a convenient device to simplify the notation. Suppose that
$$|f(x)| \leqslant K \quad \text{for all } x.$$

A. There exist step functions ϕ, ψ such that
$$\phi \leqslant f \leqslant \psi$$
(e.g. $\phi = -K\chi_{[a,b]}$, $\psi = K\chi_{[a,b]}$). Proposition 3.1.3 ensures that
$$\int \phi \leqslant \int \psi.$$

The *lower Riemann integral*
$$\underline{\int}_a^b f(x)\, dx$$
is defined to be the supremum (least upper bound) of $\int \phi$ for all such ϕ, and the *upper Riemann integral*
$$\overline{\int}_a^b f(x)\, dx$$
is the infimum (greatest lower bound) of $\int \psi$ for all such ψ. If these lower and upper integrals are equal then f is said to be *Riemann integrable* (*on* $[a, b]$) and their common value is written
$$(R)\int_a^b f(x)\, dx.$$

(i) Show that $\quad \underline{\int}_a^b f(x)\, dx \leqslant \overline{\int}_a^b f(x)\, dx.$

(ii) Show that $$(R)\int_a^b f(x)\,dx = \int_a^b f(x)\,dx$$
as previously defined when f is a step function.

(iii) Show that f is Riemann integrable if and only if there are step functions ϕ, ψ satisfying
$$\phi \leq f \leq \psi$$
with
$$\int (\psi - \phi)$$
arbitrarily small.

B. Suppose that $\{\phi_n\}$ is an increasing sequence of step functions and $\{\psi_n\}$ is a decreasing sequence of step functions where
$$\phi_n \leq f \leq \psi_n$$
for all n and
$$\int (\psi_n - \phi_n) \to 0.$$

Show that $\{\int \phi_n\}$ and $\{\int \psi_n\}$ converge to the same limit l, say, that f is Riemann integrable on $[a, b]$ and
$$(R)\int_a^b f(x)\,dx = l.$$

C. Suppose that the interval $[a, b]$ is expressed as the union $I_1 \cup \ldots \cup I_r$ of (finitely many) disjoint intervals $I_1, \ldots, I_r$. We call the set
$$P = \{I_1, \ldots, I_r\}$$
a *partition* of $[a, b]$ and refer to the greatest of the lengths $l(I_i)$ for
$$i = 1, \ldots, r$$
as the *norm* of P.

Given the partition P; we define approximating step functions for f as follows. Let
$$c_i = \inf\{f(x) : x \in I_i\},$$
$$d_i = \sup\{f(x) : x \in I_i\}$$
(remember that f is bounded), and define
$$\phi = c_1 \chi_{I_1} + \ldots + c_r \chi_{I_r},$$
$$\psi = d_1 \chi_{I_1} + \ldots + d_r \chi_{I_r}.$$
Then clearly
$$\phi \leq f \leq \psi.$$

Darboux's Theorem. *Let $\{P_n\}$ be a sequence of partitions of $[a, b]$ for which the norms converge to zero and let ϕ_n, ψ_n be the corresponding step functions approximating to f. Then*
$$\int \phi_n \to \underline{\int_a^b} f(x)\,dx,$$
$$\int \psi_n \to \overline{\int_a^b} f(x)\,dx.$$

Prove Darboux's Theorem using the following steps.

(i) Given $\epsilon > 0$; find a partition P and a corresponding step function $\phi \leq f$ such that
$$\int \phi > \underline{\int_a^b} f(x)\,dx - \epsilon/2.$$

(ii) Suppose that ϕ corresponds to the partition $P = \{I_1, \ldots, I_r\}$ of $[a, b]$ and that P_n has norm ν_n. Show that
$$\int \phi_n \geq \int \phi - (r-1) 2K\nu_n.$$

(iii) Find N such that $\int \phi_n > \underline{\int_a^b} f(x)\,dx - \epsilon$ whenever $n \geq N$.

D. A partition P' of $[a, b]$ *refines* a partition P of $[a, b]$ if every interval of P' is contained in an interval of P. Assume that P_{n+1} refines P_n for $n = 1, 2, \ldots$ in Darboux's Theorem and that f is Riemann integrable: show that $\{\phi_n\}$, $\{\psi_n\}$ then satisfy the conditions of Ex. B.

E. Given an arbitrary point p of **R**; suppose that $\{I_n\}$ is a decreasing sequence of (bounded) intervals each containing p as an interior point (i.e. not an end point) and that $l(I_n) \to 0$. Define
$$g(p) = \lim (\inf\{f(x): x \in I_n\}),$$
$$h(p) = \lim (\sup\{f(x): x \in I_n\}).$$

(i) Show that these limits exist and are independent of the choice of sequence $\{I_n\}$ converging to the point p.
(ii) Show that
$$g \leq f \leq h.$$
(iii) Show that f is continuous at the point p if and only if
$$g(p) = h(p).$$
(iv) Find g, h in the cases where $f = \chi_{[a, b]}$; $f = \chi_{[a, b)}$; $f = \chi_S$ (S is the set of rational points in $[a, b]$).

F. **Theorem.** *The function $f: \mathbf{R} \to \mathbf{R}$ is Riemann integrable on $[a, b]$ if and only if f is bounded on $[a, b]$ and the points of discontinuity of f on $[a, b]$ form a null set. In this case*
$$(R)\int_a^b f(x)\,dx = \int_a^b f(x)\,dx. \quad (\text{Lebesgue's})$$

Prove this theorem using the following steps.
(i) Define increasing and decreasing sequences of step functions $\{\phi_n\}$, $\{\psi_n\}$ by successive halving of the interval $[a, b)$ as in the proof of Theorem 1. Give the value $f(b)$ to ϕ_n and ψ_n at the point b to ensure that
$$\phi_n \leq f \leq \psi_n.$$

Darboux's Theorem now shows that

$$\int \phi_n \to \underline{\int}_a^b f(x)\,dx,$$

$$\int \psi_n \to \overline{\int}_a^b f(x)\,dx.$$

(ii) Define g, h as in Ex. E and show that

$$g(p) \leq \lim \phi_n(p) \leq \lim \psi_n(p) \leq h(p)$$

for all points p. If p is not one of the (countably many) division points in the successive halving, show that

$$g(p) = \lim \phi_n(p),$$
$$h(p) = \lim \psi_n(p).$$

(iii) Assume that the points of discontinuity of f form a null set. By Exercise E, $f = g = h$ almost everywhere. Use (ii) above to show that $f \in L^{\text{inc}}$, $f \in L^{\text{dec}}$ and hence that

$$\underline{\int}_a^b f(x)\,dx = \overline{\int}_a^b f(x)\,dx = \int_a^b f(x)\,dx.$$

(We write L^{dec} in the obvious way for $-L^{\text{inc}}$.)

(iv) Assume that

$$\underline{\int}_a^b f(x)\,dx = \overline{\int}_a^b f(x)\,dx = l,$$

say. Use (ii) above to show that $g \in L^{\text{inc}}$, $h \in L^{\text{dec}}$ and

$$\int g = \int h = l.$$

Deduce from Theorem 5.1.1, Corollary that

$$f = g = h$$

almost everywhere, and $\qquad \int f = l.$

[Recall from p. 47 that the 'if' part of this theorem may be deduced from the proof of Theorem 1 using only Ex. B above.]

3.4 Relation to the 'indefinite integral'

Most people first meet integration as the 'inverse' operation, in a loose sense, of differentiation. From the fact that the derivative of $x^3/3$ is x^2 we deduce that the 'indefinite integral' of x^2 is $x^3/3$ plus a constant. And then if we want to evaluate the 'definite integral of x^2 from 1 to 2'

3.4] INDEFINITE INTEGRAL

we learn to evaluate $x^3/3$ at the end points and subtract. In symbols:

$$\int_1^2 x^2 dx = [x^3/3]_1^2$$
$$= 2^3/3 - 1^3/3$$
$$= 7/3.$$

In the present section we propose to show how this idea of integration relates to the Lebesgue integral as we have defined it. We must obviously assume a rudimentary knowledge of differentiation, but we shall try to restrict this to a statement of the definition and a few examples of the derivatives of well known functions.

Suppose that $f \in L^1[a, b]$. Then, for any point x in $[a, b]$, $f \in L^1[a, x]$ (see Exx. 3.2.6, 7). We define a new function F_0 by

$$F_0(x) = \int_a^x f \quad \text{for } x \text{ in } [a, b]$$
$$= 0 \quad \text{for } x \text{ outside } [a, b].$$

Theorem 1. *If f is continuous at a point p of the interval $[a, b]$, then*

$$F_0'(p) = f(p).$$

Here $F_0'(p)$ stands, as always, for the derivative of F_0 at p, i.e. the limit of
$$\frac{F_0(p+h) - F_0(p)}{h}$$

as h tends to 0, provided that this limit exists. (The definition of what is meant by this limit will appear in the course of the proof.) Continuity of f at the point a would mean continuity from the right and $F_0'(a)$ would be the *right hand derivative*, i.e. the limit of the above quotient as h tends to 0 through positive values. Similarly at b, continuity would be from the left and $F_0'(b)$ would be the *left hand derivative*.

Proof. Suppose first of all that f is continuous at a point p of the interval $[a, b)$. Given $\epsilon > 0$ we can find $\delta > 0$ small enough to ensure that $[p, p+\delta)$ is contained in $[a, b)$ and also that

$$f(p) - \epsilon < f(x) < f(p) + \epsilon$$

for all x in $[p, p+\delta)$. If $0 < h < \delta$ we have

$$F_0(p+h) - F_0(p) = \int_a^{p+h} f - \int_a^p f = \int_p^{p+h} f$$

3-2

(using Proposition 3.2.1). The above inequality and Proposition 3.2.2 give
$$h(f(p)-\epsilon) \leqslant \int_p^{p+h} f \leqslant h(f(p)+\epsilon).$$

To summarise: given $\epsilon > 0$; there exists a $\delta > 0$ such that
$$\left| \frac{F_0(p+h)-F_0(p)}{h} - f(p) \right| \leqslant \epsilon$$
provided $0 < h < \delta$. This is exactly what is meant by saying that
$$\frac{F_0(p+h)-F_0(p)}{h}$$
tends to the limit $f(p)$ as h tends to 0 through positive values.

Suppose now that p is in $(a,b]$ and
$$f(p)-\epsilon < f(x) < f(p)+\epsilon$$
for all x in $(p-\delta, p] \subset (a,b]$. For $-\delta < h < 0$ we have
$$F_0(p) - F_0(p+h) = \int_a^p f - \int_a^{p+h} f = \int_{p+h}^p f.$$

Then
$$-h(f(p)-\epsilon) \leqslant \int_{p+h}^p f \leqslant -h(f(p)+\epsilon)$$

and so
$$f(p)-\epsilon \leqslant \frac{F_0(p) - F_0(p+h)}{-h} \leqslant f(p)+\epsilon$$

(dividing by the strictly positive number $-h$). This gives
$$\left| \frac{F_0(p+h)-F_0(p)}{h} - f(p) \right| \leqslant \epsilon$$
provided $-\delta < h < 0$. Thus
$$\frac{F_0(p+h)-F_0(p)}{h}$$
tends to $f(p)$ as h tends to 0 through negative values.

Combining these two results gives $F_0'(p) = f(p)$ (with the understanding about right and left hand derivatives at a, b, respectively).

We can now prove the result we have been looking for: the so-called *Fundamental Theorem of the Calculus*.

3.4] INDEFINITE INTEGRAL

Theorem 2. *If the function F has the continuous derivative f on the closed interval $[a,b]$, then*
$$\int_a^b f(x)\,dx = F(b) - F(a).$$

The above condition means that $F'(x) = f(x)$ in $[a,b]$ where the derivatives at a and b are interpreted as right and left hand derivatives, respectively, and that f is continuous on $[a,b]$.

Proof. Define the function F_0 as above. Then $F_0'(x) = f(x)$ for every x in $[a,b]$ and so $F_0'(x) - F'(x) = 0$ for every x in $[a,b]$. Using the Mean Value Theorem (proved in the Appendix, Theorem 11) we see that $F_0 - F$ is constant on $[a,b]$. In particular
$$F_0(a) - F(a) = F_0(b) - F(b)$$
and so
$$F(b) - F(a) = F_0(b) \quad (\text{since } F_0(a) = 0)$$
$$= \int_a^b f.$$

Despite the stringent condition in Theorem 2 that f should be continuous, it is fair to say that nearly all practical integration of standard functions is based on this result (combined with certain convergence theorems which allow for integration of unbounded functions on the whole of **R**: see Chapter 5).

There is a simple and extremely useful convention which is suggested by Theorem 2:

if $a < b$ and $f \in L^1[a,b]$ then
$$\int_b^a f = -\int_a^b f.$$

(See Ex. 3.2.9.) To illustrate this convention, suppose that f is integrable on every bounded interval I on the real line **R**. (For example, f might be a step function or a function continuous at every point of **R**.) Then we may define a function F by the equation
$$F(x) = \int_0^x f + C \quad \text{for all } x \text{ in } \mathbf{R}.$$

(The arbitrary constant C is added to avoid the artificial restriction that $F(0) = 0$.) We may now prove that
$$F'(p) = f(p)$$
for any point p at which f is continuous.

If $p > 0$ this is immediate by Theorem 1; if $p \leq 0$ we choose a point $c < p$ and write
$$F(x) = \int_c^x f + \int_0^c f + C$$
from which the result then follows. Note that there is no restriction to left or right hand derivatives.

It is appropriate here to prove two of the well known results of the calculus as they arose historically in the setting of 'indefinite integration'. They are immediate consequences of basic rules of differentiation, viz. the rules for differentiating the product of two functions and the composite of two functions. (At a later stage we shall prove two much more general results without any reference to differentiation: Propositions 5.1.2, 3.)

Proposition 1 (*Integration by Parts*). *If the functions F, G have continuous derivatives f, g, respectively on $[a, b]$, then*
$$\int_a^b Fg = [FG]_a^b - \int_a^b fG.$$

(Here $[FG]_a^b$ stands for $F(b)G(b) - F(a)G(a)$.)

Proof. We write H for the product function FG. By the standard rule for differentiating a product
$$H'(x) = F(x)G'(x) + F'(x)G(x)$$
$$= F(x)g(x) + f(x)G(x)$$
for all x in $[a, b]$. But now integrating from a to b and using Theorem 2:
$$H(b) - H(a) = \int_a^b Fg + \int_a^b fG$$
which is equivalent to the required formula.

We have tacitly assumed that the functions F, G are continuous – they even have continuous derivatives f, g – and also that the products Fg, fG of pairs of continuous functions are continuous (see Ex. 3.3.1). The integrability of these products now follows from Corollary 1 of Theorem 3.3.1'.

The second formula is most readily motivated by the classical notation: as an example, suppose that we are asked to evaluate
$$\int_0^1 (1 + x^2)^{-1} dx.$$

3.4] INDEFINITE INTEGRAL

We substitute hopefully $\quad x = \tan\theta,$

$$dx = \sec^2\theta\, d\theta$$

(whatever this means!) and

$$1 + x^2 = \sec^2\theta,$$

to get $\quad \displaystyle\int_0^{\frac{1}{4}\pi} \frac{\sec^2\theta}{\sec^2\theta}\, d\theta = \int_0^{\frac{1}{4}\pi} d\theta = [\theta]_0^{\frac{1}{4}\pi} = \frac{1}{4}\pi.$

The 'limits of integration' $0, \frac{1}{4}\pi$ for θ are taken because they correspond to the limits $0, 1$ for x. The success of this method of substitution is dependent upon experience as to what substitution will simplify a given integral. Of course the same substitution, or 'change of variable' as it is often called, may be used to evaluate a whole class of integrals. For example, the substitution $x = \tan\theta$ is likely to help in the evaluation of integrals of the type

$$\int_0^1 (1+x^2)^{-k}\, dx \quad (k = 1, 2, \ldots).$$

We consider the most simple-minded substitutions. Let G have a *positive continuous derivative on the interval* $[c, d]$. Then G is increasing and the values $G(t)$ for t in $[c, d]$ form a closed interval $[a, b]$ where $a = G(c)$, $b = G(d)$ (cf. Appendix, Theorem 3). Of course, if G has a *negative* continuous derivative on $[c, d]$ we can easily replace G by $-G$.

Proposition 2 (*Integration by Substitution*). *Suppose that G has a positive continuous derivative on the closed interval $[c, d]$ and write $a = G(c), b = G(d)$. Then any function f which is continuous on the closed interval $[a, b]$ satisfies*

$$\int_a^b f(x)\, dx = \int_c^d f(G(t))\, G'(t)\, dt. \tag{1}$$

Proof. Write $\quad\displaystyle F_0(x) = \int_a^x f$

as usual, for x in $[a, b]$, and let

$$H(t) = F_0(G(t))$$

for t in $[c, d]$. The standard rule for differentiating a 'function of a function' gives
$$H'(t) = F_0'(G(t))\, G'(t)$$

and so by Theorem 1 $\quad H'(t) = f(G(t))\, G'(t)$

for t in $[c, d]$. By Theorem 2 we have at once

$$H(d) - H(c) = \int_c^d f(G(t))\, G'(t)\, dt$$

and by definition $H(d) - H(c) = F_0(b) - F_0(a)$;

as $F_0(a) = 0$ this finally gives equation (1).

Here again the integrability of the composite function $f \circ G$ and of the product $(f \circ G)\, G'$ follow because these functions are all continuous on $[c, d]$ (see Exx. 3.3.1, 5).

The formula for integration by substitution fits very well with the classical notation

$$\int_a^b f(x)\, dx.$$

This is no coincidence! The integral sign $\int$ is an elongated S standing for 'sum' and the notation

$$\int_a^b f(x)\, dx$$

is suggested by the approximating sum

$$\Sigma f(x_n)\, \delta x_n,$$

where $\quad a = x_0 < x_1 < \ldots < x_r = b$

and $\quad \delta x_n = x_n - x_{n-1}.$

This sum is just another way of writing $\int \phi$ where ϕ is the step function illustrated in Fig. 24. Now if $x = G(t)$ for t in $[c, d]$, δx is approximately equal to $G'(t)\, \delta t$ and so another approximating sum is

$$\Sigma f(G(t_n))\, G'(t_n)\, \delta t_n.$$

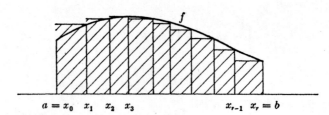

Fig. 24

Of course, this does not constitute a proof, but in Chapter 5 we shall use the convergence theorems to prove a generalisation of Proposition 2 and the proof there will show more clearly the connection with integrals of approximating step functions.

Another very good reason for using the classical notation is that it provides an expression for $\int_a^b f$ when the function f is defined by an explicit formula. For example if

$$f: \mathbf{R} \to \mathbf{R}$$

is defined by $\qquad f(x) = x^2$

for all x in $\mathbf{R}$, we write $\qquad \int_1^2 f = \int_1^2 x^2 dx.$

In this situation x is a 'dummy variable' and can be replaced by any other convenient symbol. Thus

$$\int_1^2 x^2 dx$$

is the same as $\qquad \int_1^2 t^2 dt$

or $\qquad \int_1^2 y^2 dy.$

Exercises

1. Let $I = (0, 1)$; $\qquad f(x) = \int_0^x \chi_I \quad (x \in \mathbf{R}),$

$$F(x) = \int_0^x f \quad (x \in \mathbf{R}).$$

Draw the graphs of χ_I, f and F on the same diagram.

2. If ϕ is a step function and

$$f(x) = \int_0^x \phi \quad (x \in \mathbf{R}),$$

what kind of function is f?

3. Write $\qquad \log x = \int_1^x \frac{1}{t} dt \quad$ for $\quad x > 0.$

(The log function is an exception to our convention at the beginning of this chapter: we do not extend its domain of definition to the whole of $\mathbf{R}$.) Use Proposition 2 to show that

$$\log xy = \log x + \log y \quad (x, y > 0).$$

4. Evaluate $\qquad \int_1^2 \log x \, dx.$

(See Ex. 3.)

5. Let $a_1, a_2, \ldots, a_n$ be strictly positive real numbers and define their *arithmetic mean*
$$A = \frac{1}{n}(a_1 + a_2 + \ldots + a_n)$$
and their *geometric mean* $G = \sqrt[n]{(a_1 a_2 \ldots a_n)}$.

Use the inequality $\log \dfrac{a_i}{A} \leq \dfrac{a_i}{A} - 1$

(Ex. 3.3.3) to prove that $G \leq A$.

6. If f is continuous on $[a, b]$ $(a < b)$ and k is a non-zero real number, show that
$$\int_a^b f(t)\,dt = \int_{a-k}^{b-k} f(x+k)\,dx,$$
$$\int_a^b f(t)\,dt = k \int_{a/k}^{b/k} f(kx)\,dx.$$
(Cf. Ex. 3.2.10.)

7. Prove the following result (cf. Proposition 2 and Proof).

Suppose that G has continuous derivative on $[c, d]$ and write $a = G(c)$, $b = G(d)$. If f is continuous on an interval I containing every point $x = G(t)$ $(t \in [c, d])$, then
$$\int_a^b f(x)\,dx = \int_c^d f(G(t))\,G'(t)\,dt. \qquad (1)$$

(In fact the set of all points $x = G(t)$ for t in $[c, d]$ is a bounded closed interval on **R** and so we may take I to be this interval. Cf. Appendix, Theorems 3, 9.)

Use this result to show that
$$\int_0^\pi \sin^2 \theta \cos \theta \, d\theta = 0.$$

8. Comment on the use of Proposition 2 or Ex. 7 in the following 'evaluations' of
$$\int_0^1 \sqrt{(1-x^2)}\,dx.$$

(i) The substitution $x = \sin t$ ($c = 0, d = \tfrac{1}{2}\pi$) gives
$$\int_0^{\frac{1}{2}\pi} \sqrt{(1 - \sin^2 t)} \cos t \, dt$$
$$= \int_0^{\frac{1}{2}\pi} \cos^2 t \, dt$$
$$= \int_0^{\frac{1}{2}\pi} \tfrac{1}{2}(1 + \cos 2t)\,dt$$
$$= [\tfrac{1}{2}(t + \tfrac{1}{2}\sin 2t)]_0^{\frac{1}{2}\pi}$$
$$= \tfrac{1}{4}\pi.$$

3.4] INDEFINITE INTEGRAL 63

(ii) The substitution $x = \sin t$ ($c = \pi, d = \tfrac{1}{2}\pi$) gives

$$\int_\pi^{\frac{1}{2}\pi} \sqrt{(1-\sin^2 t)} \cos t\, dt$$
$$= -\int_{\frac{1}{2}\pi}^{\pi} \cos^2 t\, dt$$
$$= [-\tfrac{1}{2}(t+\tfrac{1}{2}\sin 2t)]_{\frac{1}{2}\pi}^{\pi}$$
$$= -\tfrac{1}{4}\pi.$$

(iii) The substitution $x = \sin t$ ($c = 0, d = \tfrac{5}{2}\pi$) gives

$$\int_0^{\frac{5}{2}\pi} \sqrt{(1-\sin^2 t)} \cos t\, dt$$
$$= [\tfrac{1}{2}(t+\tfrac{1}{2}\sin 2t)]_0^{\frac{5}{2}\pi}$$
$$= \tfrac{5}{4}\pi.$$

3.5 Some further results

We cannot close this chapter without some reference to the deeper theorems which connect differentiation and integration in the Lebesgue theory. The few paragraphs that follow will give some idea of the general results that are 'well known', but we shall refer to classical text books for the proofs. The beginner should skip this section and go on to the next chapter.

Let us look first of all at Theorem 3.4.1. In what way could we hope to generalise this result? If we assume that f is continuous on the whole interval $[a, b]$, then we have $F_0'(x) = f(x)$ for all x in $[a, b]$ (with suitable interpretation about right or left hand derivatives at the end points a, b). Now we may 'interfere' with the function f at the points of a null set without altering the integral in any way. We shall obtain the same function F_0 as before, but the equation $F_0'(x) = f(x)$ will now hold only at the points of $[a, b]$ where we have not altered the values of f. For example, we may add to f the characteristic function h of the set of rational points on $[a, b]$ to obtain a function $g = f+h$ which is discontinuous at every point of $[a, b]$ (because it oscillates between values which are near to $f(p)$ and values which are near to $f(p)+1$ in every neighbourhood of the point p in $[a, b]$) and so we cannot apply Theorem 3.4.1 directly. But Theorem 3.2.3 shows that

$$G_0(x) = \int_a^x g = \int_a^x f = F_0(x)$$

for x in $[a, b]$ and so $G_0'(x) = F_0'(x) = f(x)$ for x in $[a, b]$. Thus $G_0'(x) = g(x)$ for *almost all* x in $[a, b]$.

In this particular case the derivative $G_0'(x)$ exists for *all* x in $[a, b]$. If we take for f a step function with several jump discontinuities on $[a, b]$ then the corresponding function F_0 will be continuous on $[a, b]$ but will not have a derivative at the points of discontinuity of f (see Fig. 25).

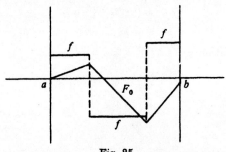

Fig. 25

One of the results that appears in almost every book on Lebesgue integration is the following.

Theorem 1. *Let $f \in L^1[a, b]$ and define F_0 by*

$$F_0(x) = \int_a^x f$$

for x in $[a, b]$. Then $F_0'(x)$ exists for almost all x in $[a, b]$ and $F_0'(x) = f(x)$ for almost all x in $[a, b]$.

In view of the simple examples we have mentioned above this is a 'best possible' result and represents a triumph for the theory. The proof requires a rather detailed discussion of 'upper and lower, right and left hand derivatives' and we do not consider this to be appropriate to a first look at Lebesgue integration.

Before we discuss any possible generalisations of Theorem 3.4.2 we must look more carefully at the function F defined by

$$F(x) = \int_a^x f + C \tag{1}$$

where we assume only that f is integrable on $[a, b]$. As usual, failing any information about f outside $[a, b]$, we may set $F(x) = 0$ for x outside $[a, b]$, but we are only interested in the behaviour of F on $[a, b]$.

First of all, and of great importance,

$$F \text{ is continuous on } [a, b]. \tag{2}$$

This is important because it shows that the process of integration

'smooths out' even the wildest functions provided only that they are integrable. If f is *bounded* on $[a,b]$ the proof is extremely simple: assume that
$$|f(x)| \leq K$$
for x in $[a,b]$ then
$$F(y) - F(x) = \int_a^y f - \int_a^x f = \int_x^y f$$
for x, y in $[a,b]$ (using the convention introduced on p. 57 if $y < x$). Thus
$$|F(y) - F(x)| \leq K|y-x| \qquad (3)$$
for x, y in $[a,b]$ by the Mean Value Theorem for Integrals (Proposition 3.2.2), and the continuity of F is immediate.

The general case is not quite so straightforward. An arbitrary f in $L^1[a,b]$ is the difference of two functions which belong to $L^{\text{inc}}[a,b]$, and the difference of two continuous functions is continuous, so we may assume without loss of generality that $f \in L^{\text{inc}}[a,b]$. There is an increasing sequence $\{\phi_n\}$ of step functions which converges to f almost everywhere in $[a,b]$ and such that
$$\int_a^b \phi_n \to \int_a^b f.$$
Given $\epsilon > 0$, we find N such that
$$0 \leq \int_a^b (f - \phi_N) < \epsilon.$$
Having chosen N (depending on ϵ) we note that ϕ_N is bounded by a constant K, say (which depends on N and ultimately on ϵ):
$$|\phi_N(x)| \leq K$$
for x in $[a,b]$. Now
$$F(y) - F(x) = \int_x^y f = \int_x^y (f - \phi_N) + \int_x^y \phi_N$$
and so
$$|F(y) - F(x)| \leq \int_a^b (f - \phi_N) + K|y - x|$$
which is less than 2ϵ provided we choose $|y - x| < \epsilon/K$. This proves that F is continuous on $[a,b]$.

The second fact that we can prove about F is easier to establish.

F is the difference of two increasing functions on $[a,b]$. (4)

(This is equivalent to saying that F has *bounded variation* on $[a,b]$: see Exx. 1–4.) To prove this we only need to express f as
$$f = f^+ - f^-,$$

where f^+, f^- are the positive and negative parts of f defined in Ex. 3.1.4. Since
$$f^+ = \tfrac{1}{2}(|f|+f),$$
$$f^- = \tfrac{1}{2}(|f|-f)$$
we may appeal to Theorem 3.2.2 to see that f^+ and f^- both belong to $L^1[a,b]$. The equation
$$\int_a^x f = \int_a^x f^+ - \int_a^x f^-$$
gives the required expression for $F(x) - F(a)$ on $[a,b]$.

We recall from the end of §3.3 that the graph of an increasing function is continuous except possibly for a countable number of jumps. If the function is also continuous we could say rather loosely that the graph is 'well behaved'. We cannot expect it to have a 'gradient' at *every* point, but we might reasonably expect it to have a gradient at *almost every* point (Fig. 26).

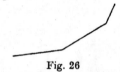

Fig. 26

One of the highlights of Lebesgue's original lectures was his proof that *an increasing continuous function has a (finite) derivative almost everywhere*. This comes at the very end of his course and uses the full power of his theory of integration (see [12]). In later years proofs appeared which were quite 'elementary' in the sense that they were from 'first principles'. A clear, though extremely subtle, exposition is given in the opening pages of Riesz–Nagy [21].

Of course, this result extends at once to the difference of two such functions. In view of this we might hope that a function F which satisfies (2) and (4) on $[a,b]$ and which therefore had a derivative f, say, almost everywhere on $[a,b]$ would satisfy
$$\int_a^b f = F(b) - F(a).$$

In fact this is not true as Ex. 10 below will show. We must assume a little more about F to be sure that the conclusion of Theorem 3.4.2 holds. The clue is in the proof we gave that F is continuous on $[a,b]$. The essential part of the proof was to show that
$$\int_I f \to 0 \quad \text{as} \quad l(I) \to 0,$$

where I is an interval in $[a, b]$. In other words, the integral of f on I can be made arbitrarily small by taking the length of I sufficiently small.

We can generalise this to the sets which we introduced at the end of §3.1 for which we have a notion of length. Almost exactly the same proof as for (2) yields the following property:

Given $\epsilon > 0$; we can find $\delta > 0$ such that

$$\left| \int_S f \right| < \epsilon \quad \text{whenever} \quad l(S) < \delta, \tag{5}$$

where S is a finite union of disjoint intervals in $[a, b]$.

This may be expressed in terms of F: given $\epsilon > 0$; there exists $\delta > 0$ such that

$$\left| \sum_{i=1}^{r} \{F(b_i) - F(a_i)\} \right| < \epsilon \quad \text{whenever} \quad \sum_{i=1}^{r} (b_i - a_i) < \delta, \tag{6}$$

where the intervals (a_i, b_i) $(i = 1, 2, ..., r)$ contained in $[a, b]$ are disjoint.

Property (6) states that F is *absolutely continuous* on $[a, b]$. The absolute continuity of F on $[a, b]$ obviously implies that F is continuous on $[a, b]$ and, not quite so obviously, that F has bounded variation on $[a, b]$ (see Ex. 8). What is far from obvious is that the absolute continuity of F is a *necessary and sufficient condition* for F to be given by

$$F(x) - F(a) = \int_a^x f \quad (x \in [a, b])$$

for some suitable f in $L^1[a, b]$. This is a particular case of the Radon–Nikodym Theorem which we shall prove in the sequel. It leads to a general 'Fundamental Theorem of the Calculus':

Theorem 2. *If F is absolutely continuous on $[a, b]$, then $F'(x) = f(x)$, say, exists for almost all x in $[a, b]$, f is integrable on $[a, b]$, and*

$$\int_a^b f = F(b) - F(a).$$

In Chapter 5, when we have some general convergence theorems at our disposal, we shall be able to prove a 'Fundamental Theorem of the Calculus' which is more general than Theorem 3.4.2 and should be enough to tackle any problems that the reader is likely to meet in practice. For proofs of Theorems 1 and 2 and a discussion of differentiation, bounded variation and absolute continuity the reader is referred to any one of the books [2], [4], [23].

Exercises

1. Let f belong to $L^1[a,b]$ and define F by equation (1). If
$$a = x_0 < x_1 < \ldots < x_r = b$$
show that
$$\sum_{i=1}^{r} |F(x_i) - F(x_{i-1})| \leq \int_a^b |f|.$$

A function F for which the sums
$$\sum_{i=1}^{r} |F(x_i) - F(x_{i-1})|$$
are bounded for all subdivisions of $[a,b]$ is said to have *bounded variation* on $[a,b]$; in this case the supremum (least upper bound) T of these sums is called the *total variation* of F on $[a,b]$.

Thus F has bounded variation on $[a,b]$ and $T \leq \int_a^b |f|$. (In Ex. 5.2.11 we shall show that $T = \int_a^b |f|$.)

2. Let $F = G - H$ where G, H are increasing on $[a,b]$. Show that F has bounded variation on $[a,b]$.

3. Assume that F has bounded variation on $[a,b]$. Let p, $-n$ denote the sum of all positive, negative terms, respectively, in
$$\sum_{i=1}^{r} (F(x_i) - F(x_{i-1})),$$
where $a = x_0 < x_1 < \ldots < x_r = b$. Denote by P, N, respectively, the least upper bounds of all such p, n and show that
$$P - N = F(b) - F(a),$$
$$P + N = T,$$
where T is the total variation of F on $[a,b]$. The upper bounds P and N are called the *positive* and *negative variations* of F on $[a,b]$. (See Ex. 1.2.2.)

4. Assume that F has bounded variation on $[a,b]$. Define $P(x)$, $N(x)$ and $T(x)$ as the positive, negative and total variations of F on $[a,x]$ for $a \leq x \leq b$. Hence show that F is the difference of two increasing functions on $[a,b]$.

Exx. 2 and 4 together show that *F has bounded variation on $[a,b]$ if and only if F can be expressed as the difference of two increasing functions on $[a,b]$*.

5. Show that condition (6) is equivalent to the following apparently stronger condition: given $\epsilon > 0$; there exists $\delta > 0$ such that
$$\sum_{i=1}^{r} |F(b_i) - F(a_i)| < \epsilon \quad \text{whenever} \quad \sum_{i=1}^{r} (b_i - a_i) < \delta,$$
where the intervals (a_i, b_i) contained in $[a,b]$ are disjoint.

6. Show that any function F which satisfies (3) has bounded variation on $[a, b]$ and is absolutely continuous on $[a, b]$. It is usual to refer to (3) as a *Lipschitz Condition*.

7. If F has derivative f on $[a, b]$ (with the usual understanding about the end points a, b: cf. Theorem 3.4.2) and if

$$|f(x)| \leq K$$

for all x in $[a, b]$ show that the Lipschitz Condition (3) is satisfied. (Use the Mean Value Theorem: we have not assumed that f is integrable!)

Show that the function sin has bounded variation on the interval $[0, 2\pi]$. Draw the graphs of sin and its positive, negative and total variation functions (as defined in Ex. 4).

8. If F is absolutely continuous on $[a, b]$, show that F is continuous on $[a, b]$ and has bounded variation on $[a, b]$ (see Ex. 5).

9. Write out the proof (indicated in the text) that the function F defined by equation (1) is absolutely continuous on $[a, b]$.

10. The reader is referred to §2.3 on Cantor's ternary set. Show that there is a *continuous increasing* function $F : [0, 1] \to [0, 1]$ such that

$$F(x) = \tfrac{1}{2} \text{ in } \tfrac{1}{3} \leq x \leq \tfrac{2}{3},$$
$$F(x) = \tfrac{1}{4} \text{ in } \tfrac{1}{9} \leq x \leq \tfrac{2}{9},$$
$$F(x) = \tfrac{3}{4} \text{ in } \tfrac{7}{9} \leq x \leq \tfrac{8}{9},$$
$$F(x) = \tfrac{1}{8} \text{ in } \tfrac{1}{27} \leq x \leq \tfrac{2}{27},$$
$$F(x) = \tfrac{3}{8} \text{ in } \tfrac{7}{27} \leq x \leq \tfrac{8}{27},$$
$$\ldots$$

and so on (repeatedly trisecting the horizontal interval and bisecting the vertical interval).

If f is the derivative of F (almost everywhere) show that

$$\int_0^1 f(x)\, dx \neq F(1) - F(0).$$

Show, without reference to Theorem 2, that F is not absolutely continuous on $[0, 1]$.

$$f(x) = \sum_1^N \frac{b_n}{2^n} \quad \text{where } b_n = \begin{cases} a_n/2 & n < N \\ 1 & n = N \end{cases}$$

and N = smallest value $\ni a_N = 1$ or ∞ if $a_n \neq 1 \ \forall n$

$$x = \sum_1^\infty \frac{a_n}{3^n}$$

4
THE LEBESGUE INTEGRAL ON $\mathbf{R}^k$

In this chapter we extend the idea of integration to Euclidean spaces of higher dimensions. The reader will miss practically nothing if he thinks of k-dimensional space as the real line $\mathbf{R}$ (i.e. $\mathbf{R}^1$), the Euclidean plane $\mathbf{R}^2$, or the Euclidean space $\mathbf{R}^3$. After our rather detailed discussion of step functions on $\mathbf{R}$ (given in §3.1) it is easy to define step functions on $\mathbf{R}^2$ and $\mathbf{R}^3$ using the characteristic functions of intervals which are now 'rectangles' or 'rectangular boxes' with sides parallel to the given 'axes'. The integrals of these step functions are defined almost exactly as before and then the extension made to functions which are limits almost everywhere of increasing sequences of step functions.

The famous Theorem of Fubini is proved almost as soon as these ideas are formulated. This makes it possible to see at once how the intuitive ideas of length, area and volume are related and also shows how integrals on $\mathbf{R}^2$ or $\mathbf{R}^3$ can be evaluated (at least in theory) by considering repeated integrals on the line $\mathbf{R}$.

4.1 Step functions on $\mathbf{R}^k$

We consider the set $\mathbf{R}^k$ which consists of rows $(x_1, x_2, ..., x_k)$ where $x_1, x_2, ..., x_k$ are real numbers. The row $(x_1, x_2, ..., x_k)$ is written x for short and called the *point* x. The distance between points x and y is defined to be
$$|x-y| = \{(x_1-y_1)^2 + ... + (x_k-y_k)^2\}^{\frac{1}{2}}.$$

The distance $|x|$ between x and 0 is called the *Euclidean norm* of x. With this distance, $\mathbf{R}^k$ is called k-*dimensional Euclidean space*; $\mathbf{R}^1$ is just the same as the real line $\mathbf{R}$ and $\mathbf{R}^2$, $\mathbf{R}^3$ are the familiar Euclidean plane and Euclidean space of three dimensions. More accurately, $\mathbf{R}^k$ is a model of k-dimensional Euclidean space in which a particular 'frame of reference' consisting of the origin $(0, 0, ..., 0)$ and the k points $(1, 0, ..., 0)$, $(0, 1, 0, ..., 0)$, $(0, 0, ..., 0, 1)$ is given a favoured place.

If $a_i \leqslant b_i$ for $i = 1, 2, ..., k$ the set of points x in $\mathbf{R}^k$ which satisfy the inequalities
$$a_i \prec x_i \prec b_i$$
for $i = 1, 2, ..., k$ form a *bounded interval* I. We write $\prec$ for either

⩽ or <; this notation saves an intolerable listing of all the kinds of intervals – there were 4 in $\mathbf{R}^1$ (of types [,], (,), [,), (,]) and there will be 4^k in $\mathbf{R}^k$. The interval I is *closed* if all the inequalities are ⩽ and is *open* if all the inequalities are <. Provided I is not empty, we may recapture a_i, b_i as infima and suprema, just as for $\mathbf{R}^1$, and we refer to the closed interval $\bar{I}$ defined by the inequalities

$$a_i \leqslant x_i \leqslant b_i$$

as the *closure* of I in this case. A bounded interval with all its sides of equal length is called a *cube*.

In $\mathbf{R}^1$ we gave a *length* to the interval I, viz. $b_1 - a_1$. In $\mathbf{R}^2$ the *area* of I is $(b_1 - a_1)(b_2 - a_2)$ and in $\mathbf{R}^3$ the *volume* of I is $(b_1 - a_1)(b_2 - a_2)(b_3 - a_3)$. To get a comprehensive term for $\mathbf{R}^k$ which includes all of these we define the *measure* of I to be

$$m(I) = (b_1 - a_1)(b_2 - a_2) \ldots (b_k - a_k).$$

Note that, according to this definition, a face or edge of a bounded interval in $\mathbf{R}^3$ has zero volume because at least one of the factors in the above expression for $m(I)$ is zero. More generally any 'face' of a bounded interval in $\mathbf{R}^k$ of dimension less than k has k-dimensional measure zero.

If I is defined by the inequalities

$$a_i < x_i < b_i$$

for $i = 1, 2, \ldots, k$, and if I is not empty, we refer to the $2k$ hyperplanes

$$x_i = a_i, \quad x_i = b_i \quad (i = 1, 2, \ldots, k)$$

as the *bounding hyperplanes* of I. They are analogous to the end points of a bounded interval in $\mathbf{R}^1$. These bounding hyperplanes are parallel to the hyperplanes $x_i = 0$ of the favoured frame of reference. In Chapter 6 we shall consider transformations of $\mathbf{R}^k$ including 'rigid motions' like rotations and translations and show that the ideas of area, volume and measure which we develop are truly 'geometric' in the sense that they do not depend on the position of the frame of reference.

We shall hardly ever use *unbounded intervals* in $\mathbf{R}^k$, but they are defined, as one would expect from the one-dimensional case, by some inequalities of the above type

$$a_i < x_i < b_i$$

and some inequalities like $\quad a_j < x_j$

or $\quad x_j < b_j.$

We are going to study real valued functions f defined on $\mathbf{R}^k$; in symbols
$$f: \mathbf{R}^k \to \mathbf{R}.$$

As before, a function which is defined on part of $\mathbf{R}^k$, is extended, if necessary, to the whole of $\mathbf{R}^k$ by giving it the value zero outside the original domain of definition.

Our task in this chapter is to define the integral $\int f$ for as many as possible of the functions $f: \mathbf{R}^k \to \mathbf{R}$ and to make sure, among other things, that $\int$ is a *linear operator*. As natural generalisations of our previous work we have the following definitions.

If S is any set of points in $\mathbf{R}^k$ we denote by χ_S the function which takes the value 1 at the points of S and the value 0 at all other points of $\mathbf{R}^k$:

$$\chi_S(x) = 1 \quad \text{if} \quad x \in S,$$
$$= 0 \quad \text{if} \quad x \notin S.$$

χ_S is called the *characteristic function* of S.

A *step function* on $\mathbf{R}^k$ is a (finite) linear combination of characteristic functions of bounded intervals:

$$\phi = c_1 \chi_{I_1} + \ldots + c_r \chi_{I_r},$$

where $I_1, \ldots, I_r$ are bounded intervals in $\mathbf{R}^k$ and $c_1, \ldots, c_r$ are real numbers.

We check at once that *the step functions on $\mathbf{R}^k$ form a linear space over the real numbers*. As in the one-dimensional case we have the following fundamental property.

Proposition 1. *Any step function may be expressed as a (finite) linear combination of characteristic functions of disjoint intervals.*

Proof. The step function

$$\phi = c_1 \chi_{I_1} + \ldots + c_r \chi_{I_r}$$

is given. Taking a lead from the proof of Proposition 3.1.1 we assume that the intervals $I_1, \ldots, I_r$ are non-empty and use their bounding hyperplanes to divide $\mathbf{R}^k$ into disjoint intervals. If there are n_i distinct bounding hyperplanes parallel to $x_i = 0$ $(i = 1, 2, \ldots, k)$ then there are $(2n_1 + 1) \ldots (2n_k + 1)$ disjoint intervals defined in this way, of which $(2n_1 - 1) \ldots (2n_k - 1)$ are bounded. As before, each of the intervals $I_1, \ldots, I_r$ is the union of certain of these disjoint bounded intervals and hence ϕ is expressed as a linear combination of their characteristic functions.

If I is a bounded interval in $\mathbf{R}^k$ then we define
$$\int \chi_I = m(I).$$

This definition is strictly analogous to the definition given for a bounded interval in $\mathbf{R}$. In $\mathbf{R}^2$ it agrees with the following intuitive picture. The 'graph' of a function $f: \mathbf{R}^2 \to \mathbf{R}$ is the 'surface' in $\mathbf{R}^3$ consisting of the points $(x_1, x_2, f(x_1, x_2))$. If $f(x_1, x_2) \geq 0$ for all x_1, x_2 and if we think of the integral of f as the 'volume' between this surface and the 'horizontal plane' with equation $x_3 = 0$, then the integral of χ_I (defined on $\mathbf{R}^2$) is the volume of the interval with base I and height 1 (see Fig. 27).

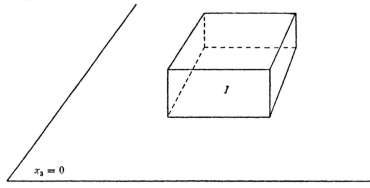

Fig. 27

In order that $\int$ should act as a linear operator on the linear space of step functions we are led to the following definition: if
$$\phi = c_1 \chi_{I_1} + \ldots + c_r \chi_{I_r}$$
then
$$\int \phi = c_1 m(I_1) + \ldots + c_r m(I_r).$$

The linearity of $\int$ is immediate, but there is the same trouble as before that the expression of ϕ as a linear combination of characteristic functions of bounded intervals is not unique, and so the consistency of the definition of $\int \phi$ has to be checked.

We consider the ambiguity in two special cases.

(i) Let I be a bounded interval in $\mathbf{R}^k$. A hyperplane $x_i = c_i$ may be used to subdivide I into three disjoint intervals
$$I = I' \cup J \cup I'',$$

where I', J, I'' consist of the points $x = (x_1, ..., x_k)$ of I which satisfy $x_i < c_i$, $x_i = c_i$, $x_i > c_i$, respectively (Fig. 28). It follows that

$$\chi_I = \chi_{I'} + \chi_J + \chi_{I''}. \tag{1}$$

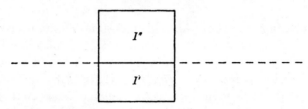

Fig. 28

As $m(J) = 0$ the consistency of the definition of integration in this case is equivalent to
$$m(I) = m(I') + m(I'')$$
which follows at once from our definition of the measure of a bounded interval in $\mathbf{R}^k$.

(ii) For any two real numbers c_1, c_2 and any bounded interval I,

$$(c_1 + c_2)\chi_I = c_1 \chi_I + c_2 \chi_I. \tag{2}$$

The consistency here amounts to the fact that

$$(c_1 + c_2) m(I) = c_1 m(I) + c_2 m(I)$$

which follows from the distributive law for real numbers.

Proposition 2. *If*

$$c_1 \chi_{I_1} + ... + c_r \chi_{I_r} = d_1 \chi_{J_1} + ... + d_s \chi_{J_s}$$

then $\quad c_1 m(I_1) + ... + c_r m(I_r) = d_1 m(J_1) + ... + d_s m(J_s).$

Proof. By the process of Proposition 1 we may use the bounding hyperplanes of all the (non-empty) intervals $I_1, ..., I_r, J_1, ..., J_s$ to obtain a common refinement
$$e_1 \chi_{K_1} + ... + e_t \chi_{K_t}$$
of both expressions

$$c_1 \chi_{I_1} + ... + c_r \chi_{I_r}, \quad d_1 \chi_{J_1} + ... + c_s \chi_{J_s}$$

in terms of disjoint intervals $K_1, ..., K_t$. This process of refinement may be carried out as a succession of simple steps of types (i) or (ii): each time an interval is subdivided we use equation (1), and then use equation (2) to gather terms involving the same interval. As the

consistency has been checked for each simple step it follows that
$$c_1 m(I_1) + \ldots + c_r m(I_r), \quad d_1 m(J_1) + \ldots + d_s m(J_s)$$
are both equal to $\quad e_1 m(K_1) + \ldots + e_t m(K_t).$

This completes the proof.

If f and g are real valued functions defined on $\mathbf{R}^k$ we write $f \geq g$ as a shorthand for $f(x) \geq g(x)$ for all x in $\mathbf{R}^k$. The proof of the following result is exactly as before.

Proposition 3. *If $\phi \geq \psi$ then $\int \phi \geq \int \psi$.*

A subset of $\mathbf{R}^k$ which is the union of finitely many disjoint bounded intervals is usually called an *elementary figure* in $\mathbf{R}^k$ (see Fig. 29).

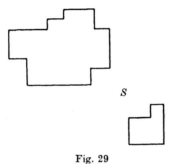

Fig. 29

Let $\mathcal{R}$ denote the collection of all the elementary figures in $\mathbf{R}^k$. As in the one-dimensional case we may prove that $S \in \mathcal{R}$ if and only if χ_S is a step function. If
$$S = I_1 \cup \ldots \cup I_r,$$
where $I_1, \ldots, I_r$ are disjoint bounded intervals, then we may safely define
$$m(S) = m(I_1) + \ldots + m(I_r).$$
There is no need to check the consistency of this definition as it is equivalent to defining
$$m(S) = \int \chi_S,$$
where $S \in \mathcal{R}$. If S, T are elementary figures then so are $S \cup T$, $S \cap T$, $S \setminus T$, $S \triangle T$, and we may verify that
$$m(S \cup T) + m(S \cap T) = m(S) + m(T)$$
(see Ex. 4). These are the first details that have to be verified in any treatment that begins by building up the so called *Lebesgue measure* in $\mathbf{R}^k$ and deducing the Lebesgue integral from it.

Exercises

1. In $\mathbf{R}^2$ the intervals I_1, I_2, I_3 are defined by the inequalities

$$I_1: 1 \leq x < 2, \quad 0 < y \leq 3,$$
$$I_2: 1 < x < 3, \quad 1 < y < 2,$$
$$I_3: 0 \leq x \leq 3, \quad 0 \leq y \leq 3.$$

Express the following step functions in terms of disjoint intervals:

$$\chi_{I_1} - \chi_{I_2}, \quad \chi_{I_1} + \chi_{I_2}, \quad \chi_{I_1} - \chi_{I_2} + \chi_{I_3}.$$

2. In $\mathbf{R}^3$ the intervals I_1, I_2 are defined by the inequalities

$$I_1: 1 \leq x < 2, \quad 0 < y \leq 3, \quad 0 < z < 2,$$
$$I_2: 1 < x < 3, \quad 1 < y < 2, \quad 1 < z < 2.$$

Express
$$7\chi_{I_1} - 5\chi_{I_2}$$

in terms of disjoint intervals.

3. Extend the notation and results of Exx. 3.1.2–4 to the case of functions from $\mathbf{R}^k$ to $\mathbf{R}$.

4. Extend the notation and results of Exx. 3.1.5–8 to the class $\mathscr{R}$ of elementary figures in $\mathbf{R}^k$.

5. Let ϕ be a step function on $\mathbf{R}^k$, a a point in $\mathbf{R}^k$ and r a non-zero real number; define

$$\psi(x) = \phi(x+a),$$
$$\theta(x) = \phi(rx)$$

for all x in $\mathbf{R}^k$. Show that ψ, θ are step functions on $\mathbf{R}^k$ and

$$\int \psi = \int \phi,$$
$$\int \theta = \frac{1}{|r|^k} \int \phi.$$

(If $x = (x_1, \ldots, x_k)$ and $r \in \mathbf{R}$ we write rx for $(rx_1, \ldots, rx_k)$.)

6. Let $I_1, \ldots, I_k$ be intervals in $\mathbf{R}^1$; the product interval

$$I = I_1 \times I_2 \times \ldots \times I_k$$

consists of all points $(x_1, x_2, \ldots, x_k)$ in $\mathbf{R}^k$ for which $x_1 \in I_1, x_2 \in I_2, \ldots, x_k \in I_k$. Show that

$$\chi_I(x) = \chi_{I_1}(x_1) \cdots \chi_{I_k}(x_k).$$

Deduce that any step function on $\mathbf{R}^k$ is a linear combination of characteristic functions of bounded *open* intervals.

4.2 The Lebesgue integral on $\mathbf{R}^k$

The theory of null sets on $\mathbf{R}$ which we introduced in Chapter 2 has an obvious extension to $\mathbf{R}^k$. A set of points in $\mathbf{R}^k$ is *null* if it can be covered by a sequence of (bounded) open intervals whose total measure is arbitrarily small. (The *total measure* of the sequence of intervals $\{I_n\}$ is the sum of the series $m(I_1) + m(I_2) + \ldots$ provided this series is convergent.)

According to this definition a *$(k-1)$-dimensional face of a bounded interval in $\mathbf{R}^k$ is null* as it can be covered by a single interval (let alone a sequence of intervals) which is arbitrarily 'thin' and so has arbitrarily small measure.

As in Chapter 2 we have

Proposition 1. *Any countable set in $\mathbf{R}^k$ is null.*

Proposition 2. *The union of a sequence of null sets in $\mathbf{R}^k$ is again a null set.*

Every step in the building up of the Lebesgue integral on $\mathbf{R}^k$ is exactly as in the previous chapter. There is so much repetition that we shall give very little further explanation in the rest of this section.

A property which holds for all points of $\mathbf{R}^k$ outside some null set is said to hold *almost everywhere* (a.e.) *in $\mathbf{R}^k$*. We shall also say that the property $\mathfrak{P}(x)$ holds for *almost all x* (a.a. x).

Theorem 1. *If $\{\phi_n\}$ is an increasing sequence of step functions on $\mathbf{R}^k$ for which the sequence $\{\int \phi_n\}$ converges, then $\{\phi_n\}$ converges almost everywhere (i.e. $\{\phi_n(x)\}$ converges for almost all x).*

Theorem 1 (Converse). *If S is a null set in $\mathbf{R}^k$, then there is an increasing sequence $\{\psi_n\}$ of step functions for which the sequence $\{\int \psi_n\}$ converges and such that $\{\psi_n(x)\}$ diverges for every x in S.*

The proofs of these results may be taken almost verbatim from the proofs in Chapter 3 – the only changes are that we replace the length $l(S)$ by the measure $m(S)$.

Suppose that $\{\phi_n\}$ is an increasing sequence of step functions on $\mathbf{R}^k$ for which $\{\int \phi_n\}$ converges and that $\phi_n(x) \to f(x)$ for almost all x, then we define the integral of f by the equation

$$\int f = \lim \int \phi_n.$$

We denote by $L^{\text{inc}}(\mathbf{R}^k)$ or L^{inc} the set of all functions f that arise in this way as limits almost everywhere of increasing sequences of step functions whose integrals are bounded.

The consistency of this definition of $\int f$ depends as before on two lemmas whose statements are true in $\mathbf{R}^k$ without any change of wording. The only changes in the proof of Lemma 1 are fairly obvious. The closed interval $[a,b]$ is replaced by a bounded closed interval I in $\mathbf{R}^k$. The set of points x at which the step function ϕ_n has jump discontinuities is a finite union of $(k-1)$-dimensional faces of bounded intervals in $\mathbf{R}^k$ and as such is null. (See the remark following the definition of null sets in $\mathbf{R}^k$.) In view of Proposition 2 the set A of points x at which there is a jump discontinuity of some ϕ_n $(n = 1, 2, \ldots)$ is therefore null.

The set of all functions f of the form $g-h$ where $g, h \in L^{\text{inc}}(\mathbf{R}^k)$ is denoted by $L^1(\mathbf{R}^k)$, or simply L^1 if the space $\mathbf{R}^k$ is understood. Further,

$$\int f = \int g - \int h.$$

To conform with classical notation we also write $\int f$ as

$$\int f(x)\, dx$$

or even

$$\int f(x_1, \ldots, x_k)\, d(x_1, \ldots, x_k).$$

But we do not repeat the integral sign when only one integration is performed. Cf. Fubini's Theorem in §4.3.

Theorem 2. (i) *If $f_1, f_2 \in L^1$ and $c_1, c_2 \in \mathbf{R}$, then $c_1 f_1 + c_2 f_2 \in L^1$ and*

$$\int (c_1 f_1 + c_2 f_2) = c_1 \int f_1 + c_2 \int f_2.$$

(ii) *If $f \in L^1$ and $f \geqslant 0$ almost everywhere, then*

$$\int f \geqslant 0.$$

(iii) *If $f \in L^1$ then $|f| \in L^1$ and*

$$\left| \int f \right| \leqslant \int |f|.$$

Theorem 2 may be summarised by saying that the Lebesgue integral is a *linear, positive, absolute* operator on the *linear space* L^1.

Theorem 3. *If $f_1 \in L^1$ and $f_2 = f_1$ almost everywhere, then $f_2 \in L^1$ and $\int f_2 = \int f_1$.*

The Lebesgue integral as we have defined it is on *the whole of* $\mathbf{R}^k$. To emphasise this we sometimes write $\int f$ as

$$\int_{\mathbf{R}^k} f$$

or

$$\int_{\mathbf{R}^k} f(x)\, dx.$$

Let I be any interval in $\mathbf{R}^k$ (not necessarily bounded) and let g be the function equal to f on I but vanishing everywhere else, i.e.

$$g = f\chi_I.$$

If $g \in L^1(\mathbf{R}^k)$ then we write

$$\int_I f = \int g$$

and refer to this as the *integral of f on I*. We also write

$$\int_I f(x)\, dx$$

for the integral of f on I. The functions f which are integrable on I form a linear space, written $L^1(I)$.

To obtain a manageable condition which is sufficient to ensure the integrability of functions on $\mathbf{R}^k$ we extend the definition of continuity using the distance

$$|x-y| = \{(x_1-y_1)^2 + \ldots + (x_k-y_k)^2\}^{\frac{1}{2}}$$

given at the beginning of this chapter.

The function $f : \mathbf{R}^k \to \mathbf{R}$ is *continuous at the point p* if, given an arbitrary $\epsilon > 0$, there is a $\delta > 0$ such that

$$|f(x)-f(p)| < \epsilon \quad \text{whenever} \quad |x-p| < \delta.$$

Roughly speaking, f is continuous at the point p if one can make sure that $f(x)$ varies very little from $f(p)$ by restricting the point x to lie inside the 'sphere' centre p with radius δ.

Suppose now that $f: \mathbf{R}^k \to \mathbf{R}$ vanishes outside a bounded interval I. Since the faces of I are all null sets in $\mathbf{R}^k$ we may assume for convenience that all the inequalities defining I are of the form

$$a_i \leqslant x_i < b_i.$$

The removal or insertion of any number of faces will have no effect on $\int f$ by Theorem 3. The interval I is now subdivided into 2^k intervals of the same type by halving (Fig. 30). For example x_i satisfies either

$$a_i \leq x_i < (a_i + b_i)/2$$

or
$$(a_i + b_i)/2 \leq x_i < b_i.$$

Call these intervals $I_1, I_2, \ldots, I_r$ $(r = 2^k)$.

If we also *assume that f is bounded*, i.e. there is a real number K such that
$$-K \leq f(x) \leq K$$

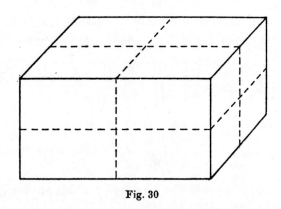

Fig. 30

for all x in $\mathbf{R}^k$, then we may define a step function

$$\phi_1 = c_1 \chi_{I_1} + \ldots + c_r \chi_{I_r},$$

where
$$c_i = \inf\{f(x) : x \in I_i\}$$
for $i = 1, 2, \ldots, r$.

By construction
$$\phi_1 \leq f \leq K\chi_I$$

and so
$$\int \phi_1 \leq Km(I).$$

If we divide every interval I_i into 2^k intervals by halving we obtain in exactly the same way a step function ϕ_2 which satisfies

$$\phi_2 \leq f$$

and
$$\int \phi_2 \leq Km(I).$$

Also
$$\phi_1 \leq \phi_2$$

since the infima can only increase as the intervals are halved. By successive halving we obtain an increasing sequence $\{\phi_n\}$ of step functions on $\mathbf{R}^k$ for which $\phi_n \leq f$ and $\{\int \phi_n\}$ is convergent. In view of

Theorem 1 we can be sure that $\{\phi_n(x)\}$ is convergent for almost all x. In order to ensure that $\{\phi_n(x)\}$ *converges to* $f(x)$ for almost all x we finally assume that f *is continuous almost everywhere in* $\mathbf{R}^k$.

Theorem 4. *Let* $f\colon \mathbf{R}^k \to \mathbf{R}$ *be a function which vanishes outside a bounded interval* I *of* $\mathbf{R}^k$. *If* f *is bounded and if the points of discontinuity of* f *form a null set in* $\mathbf{R}^k$, *then*

$$f \in L^1(\mathbf{R}^k).$$

Proof. All that remains to prove is that if f is continuous at the point p of I, then $\phi_n(p) \to f(p)$. This now follows from the definition of continuity. Given $\epsilon > 0$, there exists $\delta > 0$ such that

$$|f(x) - f(p)| < \epsilon \quad \text{whenever} \quad |x - p| < \delta.$$

Choose N so large that $\quad (\operatorname{diag} I)/2^N < \delta$

(where diag I measures the diagonal of I, i.e. the distance between two opposite corners of I). One of the 2^{kN} intervals of ϕ_N contains p and lies entirely within the distance δ of p. Thus

$$f(p) - \epsilon < f(x) < f(p) + \epsilon$$

throughout this interval. By the definition of ϕ_N using infima

$$f(p) - \epsilon \leqslant \phi_N(p) \leqslant f(p).$$

But $\{\phi_n\}$ is increasing and bounded above by f, so

$$f(p) - \epsilon \leqslant \phi_n(p) \leqslant f(p)$$

for all $n \geqslant N$. Hence $\quad \phi_n(p) \to f(p)$
as required.

Another formulation of Theorem 4 is

Theorem 4'. *Let* I *be a bounded interval in* $\mathbf{R}^k$. *If* f *is bounded on* I *and if the points of discontinuity of* f *on* I *form a null set, then* $f \in L^1(I)$.

In the Appendix we prove that a real valued function which is continuous on a *bounded closed interval* I is bounded on I. We have the following special case of Theorem 4'.

Corollary. *If* f *is continuous on the bounded closed interval* I *then* f *is integrable on* I.

Exercises

1. Show that the hyperplane in $\mathbf{R}^k$ defined by the equation $x_i = c_i$ is a null set.

2. Let S be the set of points (x,y) in $\mathbf{R}^2$ for which at least one of x,y is rational. Show that S is null.

3. Let S be the set of points (x,y) in $\mathbf{R}^2$ for which x,y are irrational and $0 < x < 1, 0 < y < 1$. Evaluate
$$\int \chi_S.$$

4. Show that the results of Exx. 3.2.3–7 are valid for $\mathbf{R}^k$.

5. If $f \in L^1(\mathbf{R}^k)$, $a \in \mathbf{R}^k$, $r \in \mathbf{R}$ ($r \neq 0$), show that
$$\int f(x+a)\,dx = \int f(x)\,dx$$
and
$$\int f(rx)\,dx = \frac{1}{|r|^k}\int f(x)\,dx.$$
(See Ex. 4.1.5.)

6. In $\mathbf{R}^2$ the *open disk*, centre 0, radius r, is the set S of all points x such that $|x| < r$.

The whole plane $\mathbf{R}^2$ is covered by disjoint unit squares defined by the inequalities
$$m \leqslant x < m+1, \quad n \leqslant y < n+1,$$
where m, n are integers. Let S_1 be the union of all of these unit squares that are contained in S. Divide each unit square into four congruent squares of the same type ($[,) \times [,)$) and let S_2 be the union of all of these squares of side $\frac{1}{2}$ that are contained in S. Continuing in this way define a sequence $\{\chi_{S_n}\}$ of step functions that converges to χ_S (everywhere). Hence show that χ_S is integrable. (We now define $\int \chi_S$ to be the *area* of S.)

7. In $\mathbf{R}^2$ the *closed disk* T, centre 0, radius r is defined by $|x| \leqslant r$. Use the construction of Ex. 6 to find a decreasing sequence of step functions $\{\chi_{T_n}\}$ which converges to χ_T (everywhere) and show that χ_T is integrable. (Call $\int \chi_T$ the *area* of T.)

8. Show that the *circle*, centre 0, radius r in $\mathbf{R}^2$ defined by $|x| = r$, is null and use Theorem 4 to deduce that the areas of S, T, defined in Exx. 6, 7 above, exist and are equal.

9. In $\mathbf{R}^3$ define the *open ball* S centre a, radius r, by the inequality $|x-a| < r$ and the *closed ball* T centre a, radius r, by the inequality
$$|x-a| \leqslant r.$$

4.2]   LEBESGUE INTEGRAL   83

Generalise the construction of Exx. 6, 7 to show that both χ_S, χ_T are integrable. (The values of these integrals give the *volumes* of S, T. We shall verify later that these volumes are equal: see Ex. 4.3.4.)

The Riemann Integral on $\mathbf{R}^k$

Extend Exx. A–F at the end of §3.3 to functions $f \colon \mathbf{R}^k \to \mathbf{R}$ which are bounded and vanish outside a bounded interval I.

4.3 Fubini's Theorem

The integral we defined in the last section is capable, in theory, of integrating a great variety of functions. But we are again faced with the practical difficulty of evaluating these integrals. There is no obvious counterpart of the 'Fundamental Theorem of the Calculus' to aid us in the higher dimensional spaces $\mathbf{R}^k$ ($k \geq 2$) but there is one result of supreme importance which relates the integral on $\mathbf{R}^k$ to successive integrations on lower dimensional spaces, and ultimately to integrations on $\mathbf{R}$. This is Fubini's Theorem. We shall state and prove Fubini's Theorem for $\mathbf{R}^2$ and extend the notation and result to $\mathbf{R}^k$ in an obvious way later.

***Theorem 1** (**Fubini's Theorem for** $\mathbf{R}^2$)*. *Suppose that* $f \in L^1(\mathbf{R}^2)$. *Then*

$$\int f(x,y)\,d(x,y) = \int \left(\int f(x,y)\,dy \right) dx.$$

This equation is to be interpreted as meaning that

$$F(x) = \int f(x,y)\,dy$$

exists for almost all x in $\mathbf{R}$, that

$$\int F(x)\,dx$$

exists and equals $\displaystyle\int f(x,y)\,d(x,y)$.

(We can avoid the classical notation for integrals if we define the function
$$f_x \colon \mathbf{R} \to \mathbf{R} \quad (x \in \mathbf{R})$$
by means of the equation
$$f_x(y) = f(x,y) \quad (y \in \mathbf{R}).$$

Then F is defined for almost all x in $\mathbf{R}$ by

$$F(x) = \int f_x$$

and Fubini's Theorem becomes

$$\int f = \int F.$$

This formulation has the air of simplicity but it seems to the author to be far less memorable than the one we have given!)

The proof we shall give of Fubini's Theorem is quite elementary in the sense that it uses only the Fundamental Theorem 4.2.1 which was required even to *define* the Lebesgue integral. The steps of the proof follow the steps in the introduction of the integral. First we prove the theorem when f is the characteristic function of a bounded interval, in which case it is well nigh obvious; then we extend by linearity to step functions; the meat of the proof is the extension to functions of L^{inc} by means of increasing sequences of step functions and then finally the extension to L^1 is trivial by linearity. The subtle part of the proof is in the third section where we have to relate the concepts 'almost everywhere in $\mathbf{R}^2$' and 'almost everywhere in $\mathbf{R}$'.

Proof. (i) Let $f = \chi_I$ where I is defined by inequalities

$$a_1 < x < b_1, \quad a_2 < y < b_2;$$

then
$$\int f = m(I) = (b_1 - a_1)(b_2 - a_2)$$

and
$$\int_{a_1}^{b_1} \left(\int_{a_2}^{b_2} 1 \, dy \right) dx = \int_{a_1}^{b_1} (b_2 - a_2) \, dx = (b_1 - a_1)(b_2 - a_2).$$

(ii) Since the integral on the left hand side and both of the integrals on the right hand side are linear operators, the result is true for step functions $f = c_1 \chi_{I_1} + \ldots + c_r \chi_{I_r}$.

Note that the function F defined (for all x) by

$$F(x) = \int f(x, y) \, dy$$

is itself a step function (on $\mathbf{R}$) in this case. (See Ex. 1.)

(iii) Suppose now that $f \in L^{\text{inc}}(\mathbf{R}^2)$, that $\{\phi_n\}$ is an increasing sequence of step functions which converges to f almost everywhere in $\mathbf{R}^2$ and that

$$\int f = \lim \int \phi_n.$$

4.3] FUBINI'S THEOREM

By part (ii) we see that $\int \phi_n = \int \Phi_n,$

where the step function Φ_n is defined (for all x in $\mathbf{R}$) by

$$\Phi_n(x) = \int \phi_n(x, y) \, dy.$$

From $\phi_n \leq \phi_{n+1}$
it follows that $\Phi_n \leq \Phi_{n+1}$

(Proposition 3.1.3). But $\{\int \phi_n\}$ is convergent (by the definition of f) and so $\{\int \Phi_n\}$ is convergent. Now Theorem 4.2.1 shows that $\{\Phi_n\}$ is convergent almost everywhere in $\mathbf{R}$.

This is where we have to be careful! Let S be the set of all points (x, y) in $\mathbf{R}^2$ for which $\{\phi_n(x, y)\}$ fails to converge to $f(x, y)$. We are given that S is null. For the moment let us assume the truth of Lemma 1 below which states that the 'cross-section'

$$S_x = \{y : (x, y) \in S\}$$

of S is null for almost all x in $\mathbf{R}$. Let us choose an x_0 in $\mathbf{R}$ for which S_{x_0} is null and also $\{\Phi_n(x_0)\}$ converges: this will be true for almost all x_0 in $\mathbf{R}$. Then

$$\phi_n(x_0, y) \to f(x_0, y)$$

for almost all y (in fact for y not in S_{x_0}) and the sequence

$$\left\{ \int \phi_n(x_0, y) \, dy \right\} = \{\Phi_n(x_0)\}$$

converges. This shows that

$$\int f(x_0, y) \, dy$$

exists and equals $\lim \Phi_n(x_0)$.

Write
$$F(x_0) = \int f(x_0, y) \, dy = \lim \Phi_n(x_0).$$

Now, finally, the increasing sequence $\{\Phi_n\}$ converges to F almost everywhere in $\mathbf{R}$ and

$$\lim \int \Phi_n = \lim \int \phi_n = \int f,$$

so that $F \in L^{\mathrm{inc}}(\mathbf{R})$ and $\int F = \int f.$

(iv) If $f = g - h$ where $g, h \in L^{\text{inc}}(\mathbf{R}^2)$ then the result follows from part (iii) by linearity.

To complete the proof we need

Lemma 1. *Let S be a null set in $\mathbf{R}^2$ and define*
$$S_x = \{y : (x, y) \in S\}.$$
(See Fig. 31.) Then S_x is null for almost all x in $\mathbf{R}$.

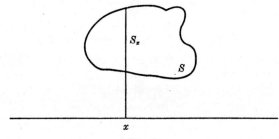

Fig. 31

Proof. By the Converse of Theorem 4.2.1 we can find an increasing sequence $\{\psi_n\}$ of step functions on $\mathbf{R}^2$ for which $\{\int \psi_n\}$ converges but $\{\psi_n(x, y)\}$ diverges for all (x, y) in S. By part (ii) of the proof of Fubini's Theorem (which, of course, did not require Lemma 1!) the step function Ψ_n defined by
$$\Psi_n(x) = \int \psi_n(x, y) \, dy$$

satisfies
$$\int \Psi_n = \int \psi_n$$

and so Theorem 4.2.1 shows that $\{\Psi_n(x)\}$ converges for almost all x in $\mathbf{R}$. Let x_0 be any such point for which $\{\Psi_n(x_0)\}$ converges, i.e. for which $\{\int \psi_n(x_0, y) \, dy\}$ converges. A final application of Theorem 4.2.1 shows that $\{\psi_n(x_0, y)\}$ converges for almost all y. But $\{\psi_n(x_0, y)\}$ diverges for all y in S_{x_0} and so S_{x_0} is null. This proves the lemma.

If we interchange the roles of x and y throughout the above proof we see that
$$\int f(x, y) \, d(x, y) = \int \left(\int f(x, y) \, dx \right) dy$$
and so we have

Corollary. *If $f \in L^1(\mathbf{R}^2)$ then*
$$\int \left(\int f(x, y) \, dx \right) dy = \int \left(\int f(x, y) \, dy \right) dx.$$

4.3] FUBINI'S THEOREM

The reader should note that this result is based on the existence of the integral of f on $\mathbf{R}^2$. At a later stage (Theorem 6.1.2) we shall prove a partial converse of Fubini's Theorem where we only assume the existence of one of the repeated integrals.

Apart from its many applications to the practical evaluation of integrals on $\mathbf{R}^2$, Fubini's Theorem sheds a great deal of light on certain theoretical questions.

We anticipate here a few thoughts from Chapter 6. Now that we have a very general definition of integration it is possible to give a very general definition of measure which agrees with our intuitive measure of intervals but which extends it to a much wider class of subsets of $\mathbf{R}^k$. If S is a subset of $\mathbf{R}^k$ and if χ_S is integrable, then

$$m(S) = \int \chi_S.$$

We recall that this is in agreement with our previous definitions when S is either a bounded interval (p. 73) or an elementary figure (p. 75). Many of the most striking properties of this measure depend on the convergence theorems of the next chapter, but Fubini's Theorem illustrates one aspect that appeals very much to the intuition.

Before we give this illustration let us remark that we may now use the words *length*, *area* and *volume* in $\mathbf{R}^1$, $\mathbf{R}^2$ and $\mathbf{R}^3$, respectively, in place of the word *measure*, and there is no longer any need to write these words under quotation marks as they have been given a precise mathematical definition.

Suppose now that S is a set of points in $\mathbf{R}^2$ and that the area $m(S) = \int \chi_S$ is defined. According to Fubini's Theorem the area of S can be calculated by the formula

$$m(S) = \int \left(\int \chi_S(x,y)\, dy \right) dx.$$

But if x is kept fixed, $\chi_S(x,y) = \chi_{S_x}(y)$ gives the characteristic function of the cross-section S_x and so

$$\int \chi_S(x,y)\, dy$$

is simply the one-dimensional measure of the cross-section S_x. Thus

$$m(S) = \int m(S_x)\, dx.$$

In other words the length of the cross-section S_x exists for almost all x and the area of S is calculated by integrating this length with respect to x (Fig. 32). This agrees with the intuitive idea of area and the familiar evaluation of areas by integration.

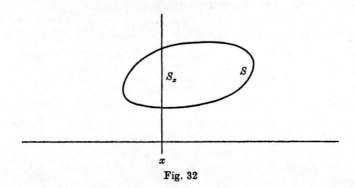

Fig. 32

As a particular case of this result we may take S to be the *ordinate set* of a given positive function $f: \mathbf{R} \to \mathbf{R}$ (see Fig. 33), viz.

$$S = \{(x,y): 0 \leqslant y \leqslant f(x)\}.$$

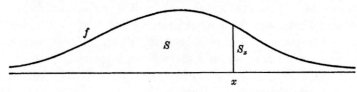

Fig. 33

In this case $m(S_x) = f(x)$. If we assume that S has an area in the above sense, then the area of S is the value of the integral

$$\int f(x)\,dx.$$

Note that we have not restricted S to lie between two ordinates $x = a$, $x = b$ although, of course, we could do so if we wanted and obtain the definite integral

$$\int_a^b f(x)\,dx.$$

(It is possible to prove the converse result that the ordinate set S of a positive function f in $L^1(\mathbf{R})$ is measurable: see Ex. 6.2.11.)

The reader may be amused to see that we can now find the area of a circular disk of radius r in $\mathbf{R}^2$. As we have pointed out at the beginning of the chapter, our definition of $\mathbf{R}^2$ depends on the favoured frame of reference and so, at first sight, the measure we have introduced may give different areas to disks of radius r if they lie in different parts of the plane! At a later stage we shall show that our measure is in fact invariant under all Euclidean 'motions' like translations and rotations (and even reflections). But the fact that our measure is invariant under translations is easy to see from first principles; this is done in Ex. 4.2.5. Let us take this for granted for the purpose of our example and assume that the centre of the disk is at the origin. The other fact that we need to check is that the disk does have an area. This follows at once from a general (quite elementary) result (Proposition 6.3.1) but we can prove it from first principles for a circular disk as in Exx. 4.2.6, 7. For definiteness we may as well include the boundary of the disk, i.e. consider a *closed* disk, though the boundary makes no difference to the area.

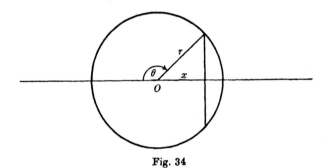

Fig. 34

According to the result deduced above from Fubini's Theorem, the area of the disk is
$$A = \int_{-r}^{r} 2\sqrt{(r^2 - x^2)}\, dx. \qquad \text{(Fig. 34)}$$

We may use the substitution $x = -r\cos\theta$ (Proposition 3.4.2) which gives
$$A = \int_{0}^{\pi} 2r\sin\theta\, r\sin\theta\, d\theta$$
$$= r^2 \int_{0}^{\pi} (1 - \cos 2\theta)\, d\theta$$
$$= r^2 \left[\theta - \frac{\sin 2\theta}{2}\right]_{0}^{\pi}$$
$$= \pi r^2.$$

The more general formulation of Fubini's Theorem for $\mathbf{R}^k$ ($k \geq 2$)

can be simplified by the introduction of a rather natural notation. Suppose that $k = l+m$ where $l \geqslant 1$, $m \geqslant 1$ and write (x, y) as shorthand for the point $(x_1, x_2, ..., x_l, y_1, y_2, ..., y_m)$ of $\mathbf{R}^k$.

Theorem 2 (Fubini's Theorem for $\mathbf{R}^k$). *Suppose that $f \in L^1(\mathbf{R}^k)$ ($k \geqslant 2$). Then*
$$\int f(x,y) \, d(x,y) = \int \left(\int f(x,y) \, dy \right) dx.$$

The interpretation of this equation is that the integral
$$\int f(x,y) \, dy$$
(on $\mathbf{R}^m$) exists for almost all x in $\mathbf{R}^l$; if we write
$$F(x) = \int f(x,y) \, dy$$
then
$$\int F(x) \, dx$$
(on $\mathbf{R}^l$) exists and equals $\int f(x,y) \, d(x,y)$ (on $\mathbf{R}^k$).

Corollary. *If $f \in L^1(\mathbf{R}^k)$ then*
$$\int \left(\int f(x,y) \, dx \right) dy = \int \left(\int f(x,y) \, dy \right) dx.$$

The proofs of these results are almost exact repetitions of the ones given above with (x,y) in $\mathbf{R}^2$ replaced by (x,y) in $\mathbf{R}^k$. The only point of difference is in the very easy first part.

Let the interval I in $\mathbf{R}^k$ be defined by the inequalities
$$a_i < x_i < b_i \quad (i = 1, 2, ..., l),$$
$$c_j < y_j < d_j \quad (j = 1, 2, ..., m)$$
and the corresponding intervals J, K in $\mathbf{R}^l$, $\mathbf{R}^m$ defined by these sets of inequalities taken separately. Then
$$\int \chi_I = m(I) = (b_1 - a_1) \ldots (b_l - a_l)(d_1 - c_1) \ldots (d_m - c_m)$$
and
$$\int_J \left(\int_K 1 \, dy \right) dx = \int_J (d_1 - c_1) \ldots (d_m - c_m) \, dx$$
$$= (b_1 - a_1) \ldots (b_l - a_l)(d_1 - c_1) \ldots (d_m - c_m).$$

4.3]   FUBINI'S THEOREM   91

The reader is encouraged to fill in the details of the rest of the proof.

We shall give one more illustration from the familiar process for evaluating a volume. Let S be a set of points in $\mathbf{R}^3$ for which the volume

$$m(S) = \int \chi_S$$

is defined. Then we can calculate this volume by integrating the areas of two-dimensional cross-sections, viz.

$$m(S) = \int m(S_x)\,dx,$$

where
(see Fig. 35).
$$S_x = \{(y, z) \in \mathbf{R}^2 : (x, y, z) \in S\}$$

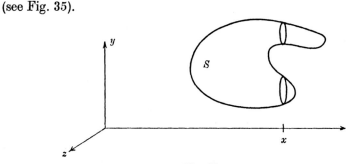

Fig. 35

If S is a (closed) ball of radius r in $\mathbf{R}^3$, then we may assume that the volume exists (Ex. 4.2.9) and that the centre is at the origin as the volume is invariant under translations (Ex. 4.2.5).

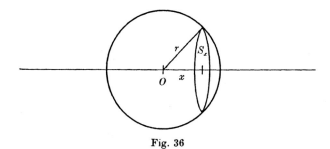

Fig. 36

The square of the radius of S_x is $r^2 - x^2$ (by our definition of distance, based on Pythagoras' Theorem) and $-r \leqslant x \leqslant r$ (see Fig. 36). The

area of S_x is therefore $\pi(r^2-x^2)$ by our previous example. The volume of the sphere is now
$$\int_{-r}^{r} \pi(r^2-x^2)\,dx = \pi[r^2x - x^3/3]_{-r}^{r}$$
$$= \pi(2r^3 - 2r^3/3)$$
$$= \tfrac{4}{3}\pi r^3.$$

A fitting point at which to end the first chapter about integration on Euclidean space!

Exercises

1. If ϕ is a step function on $\mathbf{R}^2$, verify that the function Φ defined by
$$\Phi(x) = \int \phi(x,y)\,dy \quad (x \in \mathbf{R})$$
is a step function on $\mathbf{R}$.

Generalise this to the case where ϕ is a step function on $\mathbf{R}^k$ and Φ is defined on $\mathbf{R}^l$.

2. Calculate the measure (volume) of the following sets in $\mathbf{R}^3$:
  (i) $\{(x,y,z): 0 \leqslant z \leqslant 2-x^2-y^2\}$,
  (ii) $\{(x,y,z): x+y+z \leqslant 1, x \geqslant 0, y \geqslant 0, z \geqslant 0\}$.

3. Let I denote the unit square in $\mathbf{R}^2$ defined by
$$0 < x \leqslant 1, \quad 0 < y \leqslant 1$$
and let
$$f(x,y) = \frac{x^2-y^2}{(x^2+y^2)^2} \quad \text{for } (x,y) \text{ in } I.$$
Prove that
$$\int_0^1 \left(\int_0^1 f(x,y)\,dy \right) dx = \tfrac{1}{4}\pi.$$
What is the value of the other repeated integral? Does f belong to $L^1(I)$?

4. Use Fubini's Theorem to show that the areas of the open and closed disks of Exx. 4.2.6, 7 are equal. Hence show that the volumes of the open and closed balls of Ex. 4.2.9 are equal.

5. (i) Show that the points of a straight line in $\mathbf{R}^2$ form a null set.
  (ii) Find a formula for the area of a triangle in $\mathbf{R}^2$ with a pair of sides parallel to $x=0, y=0$.
  (iii) Find a formula for the area of any (bounded) rectangle in $\mathbf{R}^2$.
  (iv) Show that areas in $\mathbf{R}^2$ are unaltered by translations, rotations and reflections.

5
THE CONVERGENCE THEOREMS

We have already stressed in Chapter 1 the importance of the property of *completeness* for the real line **R**. We saw there that any real number may be expressed as the limit of a bounded increasing sequence of rational numbers. (In fact we used decimal expansions to do this, but that is not important.) The point of the Axiom of Completeness for the real numbers is that we may now take any bounded increasing sequence of real numbers and be sure that it has a *real number* as its limit. In other words, this further taking of limits of bounded increasing sequences does not provide any more numbers.

We shall consider in an analogous way the space L^1 of Lebesgue integrable functions (over $\mathbf{R}^k$). First of all, any element of L^{inc} is the limit almost everywhere of an increasing sequence of step functions whose integrals are bounded. What if we now consider an increasing sequence $\{f_n\}$ of functions from L^{inc} whose integrals are bounded? The striking fact is that $\{f_n\}$ converges almost everywhere to a function f where f lies in L^{inc} and, as we should expect,

$$\int f = \lim \int f_n.$$

But there is a genuine difficulty. The Axiom of Completeness for **R** could equally be expressed in terms of bounded *decreasing* sequences. What does this mean for L^{inc}? On the face of it, any discussion of decreasing sequences of functions from L^{inc} would hardly be promising as the two lots of inequalities are running contrary to each other. Recall that L^1 was introduced by taking differences $f = g - h$ of functions g, h from L^{inc}. For a given f, there is some freedom in the choice of g, h; in fact we may arrange that h is positive and $\int h$ is arbitrarily small. This rescues the situation and we are able to prove a central result of the whole theory – the so called Monotone Convergence Theorem (or Beppo Levi's Theorem) – that if $\{f_n\}$ is a monotone sequence of functions from L^1 whose integrals are bounded, then $\{f_n\}$ converges almost everywhere to a function f where f lies in L^1 and where

$$\int f = \lim \int f_n.$$

All the other convergence theorems, the theory of measure and most of the applications of the Lebesgue theory to other subjects can be deduced from this result.

5.1 The Monotone Convergence Theorem

Theorem 1. *Let $\{f_n\}$ be an increasing sequence of functions in L^{inc} whose integrals are bounded. Then $\{f_n\}$ converges almost everywhere to a function f, where f lies in L^{inc} and*

$$\int f = \lim \int f_n.$$

Proof. By the definition of L^{inc} there are increasing sequences

$$\phi_{11}, \quad \phi_{12}, \quad \phi_{13}, \quad \ldots$$
$$\phi_{21}, \quad \phi_{22}, \quad \phi_{23}, \quad \ldots$$
$$\phi_{31}, \quad \phi_{32}, \quad \phi_{33}, \quad \ldots$$
$$\ldots$$

of step functions which converge almost everywhere to $f_1, f_2, f_3, \ldots$ respectively, and for which

$$\lim \int \phi_{mn} = \int f_m$$

(m fixed). Let

$$\phi_n = \max\{\phi_{ij} : 1 \leq i, j \leq n\}.$$

Then ϕ_n is a step function (see Ex. 3.1.3) and $\phi_n \leq \phi_{n+1}$ for all n. In other words $\{\phi_n\}$ is an increasing sequence of step functions.

By our construction $\phi_n \leq f_n$ almost everywhere, and we are also given

$$\int f_n \leq K$$

for all n, so that

$$\int \phi_n \leq K$$

for all n. Hence our fundamental Theorem 4.2.1 shows that $\{\phi_n\}$ converges almost everywhere to a function f of L^{inc} and

$$\int \phi_n \to \int f.$$

5.1] MONOTONE CONVERGENCE THEOREM

It only remains to show that $f_n \to f$ almost everywhere and
$$\int f_n \to \int f.$$

This we do by a 'squeezing' argument.

By the definition of ϕ_n, $\quad \phi_{mn} \leq \phi_n$

for $m \leq n$. Keeping m fixed and letting n tend to infinity,
$$f_m \leq f$$
almost everywhere, for each $m = 1, 2, \ldots$. In view of Proposition 4.2.2 this means that $\quad f_m \leq f \quad (m = 1, 2, \ldots)$

holds almost everywhere. (See the remarks after the proof.) But now
$$\phi_n \leq f_n \leq f$$
almost everywhere and $\phi_n \to f$ almost everywhere, so the sequence $\{f_n\}$ is 'squeezed' and converges to f almost everywhere. In the same way
$$\int \phi_n \leq \int f_n \leq \int f$$
and
$$\int \phi_n \to \int f$$

so that the sequence $\{\int f_n\}$ is 'squeezed' and converges to $\int f$. This completes the proof.

In the above proof we made the distinction between the two statements:

'$f_m \leq f$ almost everywhere, for each $m = 1, 2, \ldots$.'

'$f_m \leq f \quad (m = 1, 2, \ldots)$ almost everywhere.'

We shall expand this a little. Let S_m be the set of all points x in $\mathbf{R}^k$ for which $f_m(x) > f(x)$, and let S be the set of all points x in $\mathbf{R}^k$ for which $f_m(x) > f(x)$ for *some* m. Then the first statement means that S_m is null for each m and the second statement means that S is null. By the above definition S is the union of all the S_m's. Now Proposition 4.2.2 shows that the first statement implies the second. The converse is obviously true, and so the two statements are equivalent.

It is now quite clear what we mean by saying that a sequence $\{f_n\}$ is *increasing almost everywhere*. We mean that

$$f_n \leqslant f_{n+1} \text{ almost everywhere, for each } n = 1, 2, \ldots$$

or equivalently that

$$f_n \leqslant f_{n+1} \quad (n = 1, 2, \ldots) \text{ almost everywhere.}$$

The reader may like to prove that the condition of Theorem 1 that $\{f_n\}$ should be increasing may be replaced by the condition that $\{f_n\}$ should be increasing almost everywhere.

To prove the Monotone Convergence Theorem we need the following simple result.

Lemma 1. *Let $f \in L^1$. Given $\epsilon > 0$, there exist g, h in L^{inc} such that $f = g - h$, where h is positive and*

$$\int h < \epsilon.$$

Proof. By the definition of L^1, $f = g_1 - h_1$, where $g_1, h_1 \in L^{\text{inc}}$. Thus there is an increasing sequence $\{\psi_n\}$ of step functions which converges to h_1 almost everywhere and such that

$$\int \psi_n \to \int h_1.$$

Choose N so large that

$$0 \leqslant \int h_1 - \int \psi_N < \epsilon.$$

Then $h_2 = h_1 - \psi_N$ is positive almost everywhere, belongs to L^{inc} and satisfies

$$0 \leqslant \int h_2 < \epsilon.$$

Also $g_2 = g_1 - \psi_N \in L^{\text{inc}}$ and $f = g_2 - h_2$.

Finally let $h = h_2^+$ (the positive part of h_2 as defined in Ex. 3.1.4) and let $g = f + h$. As h_2 is positive almost everywhere, $h = h_2$ and $g = g_2$ almost everywhere. Thus, g, h satisfy the conditions of the lemma.

Theorem 2 (*The Monotone Convergence Theorem*). *Let $\{f_n\}$ be a monotone sequence of functions in $L^1(\mathbf{R}^k)$ whose integrals are bounded. Then $\{f_n\}$ converges almost everywhere to a function f, where f lies in $L^1(\mathbf{R}^k)$, and*

$$\int f = \lim \int f_n.$$

5.1] MONOTONE CONVERGENCE THEOREM

Proof. By considering $\{-f_n\}$, if necessary, we may assume that the sequence $\{f_n\}$ is increasing; then, by considering the sequence $\{f_n - f_1\}$, if necessary, we may assume that $f_n \geq 0$ for all n. It is convenient to use the notation of series. Let

$$a_1 = f_1, \quad a_n = f_n - f_{n-1}$$

for $n \geq 2$ so that

$$f_n = a_1 + a_2 + \ldots + a_n$$

and

$$a_n \geq 0$$

for $n \geq 1$. If we apply Lemma 1 to a_n with $\epsilon = 1/2^n$, for each $n \geq 1$, we find positive functions b_n, c_n in L^{inc} such that $a_n = b_n - c_n$ and

$$0 \leq \int c_n < 1/2^n.$$

(The positiveness of b_n follows from that of a_n and c_n as $b_n = a_n + c_n$.)

Now let

$$g_n = b_1 + b_2 + \ldots + b_n,$$
$$h_n = c_1 + c_2 + \ldots + c_n$$

so that

$$f_n = g_n - h_n,$$

where $g_n, h_n \in L^{\text{inc}}$ and the sequences $\{g_n\}, \{h_n\}$ are increasing. Moreover,

$$\int h_n < 1/2 + 1/2^2 + \ldots + 1/2^n < 1$$

and

$$\int g_n = \int f_n + \int h_n$$

are bounded. We may therefore apply Theorem 1 to see that $\{g_n\}$, $\{h_n\}$ converge almost everywhere to functions g, h of L^{inc} and hence $\{f_n\}$ converges almost everywhere to $f = g - h$ which is an element of L^1.

Finally,

$$\int g_n \to \int g, \quad \int h_n \to \int h$$

and so

$$\int f_n \to \int f.$$

As a simple, but extremely important consequence of Theorem 2 we have

Corollary. *If f is a positive element of L^1 and $\int f = 0$ then $f = 0$ almost everywhere.*

Proof. Consider the increasing sequence $\{f_n\}$ where $f_n = nf$. Then certainly the integrals $\int f_n$ are bounded – they are all zero! Hence $\{f_n\}$

converges almost everywhere. But $\{f_n(x)\}$ cannot converge unless $f(x) = 0$, and so in fact $f(x) = 0$ for almost all x.

Our first applications of the Monotone Convergence Theorem are of a very practical nature, viz. to evaluate the integrals of functions $f: \mathbf{R} \to \mathbf{R}$ which are unbounded or which do not vanish outside a bounded interval. For example, suppose that we are asked to evaluate

$$\int_{-\infty}^{\infty} e^{-|x|} \, dx$$

(see Fig. 37).

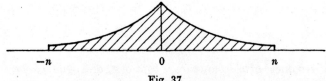

Fig. 37

We need to know that this integral exists and how to find its value. The integral

$$\int_{-n}^{n} e^{-|x|} \, dx$$

can be evaluated by means of our (elementary) Fundamental Theorem of the Calculus (Theorem 3.4.2). For

$$\int_{0}^{n} e^{-x} \, dx = [-e^{-x}]_0^n = 1 - e^{-n},$$

$$\int_{-n}^{0} e^{x} \, dx = [e^{x}]_{-n}^{0} = 1 - e^{-n}$$

and so

$$\int_{-n}^{n} e^{-|x|} \, dx = 2 - 2e^{-n}$$

(Proposition 3.4.2). From this we should like to deduce that

$$\int_{-\infty}^{\infty} e^{-|x|} \, dx = \lim (2 - 2e^{-n}) = 2.$$

The justification by means of the Monotone Convergence Theorem is immediate. We define f_n as follows:

$$f_n(x) = e^{-|x|} \quad \text{for} \quad |x| \leqslant n,$$
$$ = 0 \quad \text{for} \quad |x| > n.$$

Then $\{f_n\}$ is an increasing sequence of functions from L^1 and

$$f_n(x) \to e^{-|x|}$$

for all x in **R**. Moreover $\int f_n = \int_{-n}^{n} e^{-|x|} dx$
is bounded above by 2 and so

$$\int_{-\infty}^{\infty} e^{-|x|} dx = \lim \int f_n = 2$$

as required.

On the other hand suppose that we are asked to evaluate

$$\int_0^1 x^{-\frac{1}{2}} dx$$

(see Fig. 38).

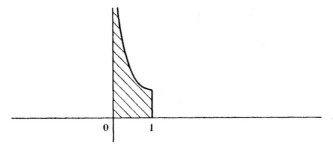

Fig. 38

The integrand $x^{-\frac{1}{2}}$ is unbounded in any interval containing the origin and so we consider the integral

$$\int_{1/n}^1 x^{-\frac{1}{2}} dx.$$

By the Fundamental Theorem of the Calculus this has value

$$[2x^{\frac{1}{2}}]_{1/n}^1 = 2(1 - n^{-\frac{1}{2}})$$

which converges to 2. To verify that

$$\int_0^1 x^{-\frac{1}{2}} dx = 2$$

we again use the Monotone Convergence Theorem, this time taking

$$f_n(x) = x^{-\frac{1}{2}} \quad \text{for} \quad 1/n \leqslant x \leqslant 1,$$
$$= 0 \quad \text{otherwise.}$$

Then $\{f_n\}$ is an increasing sequence of integrable functions such that

$$f_n(x) \to x^{-\frac{1}{2}}$$

for all x in $(0, 1]$ and the integrals $\int f_n$ are bounded above by 2. Thus

$$\int_0^1 x^{-\frac{1}{2}} dx = \lim \int f_n = 2$$

as required.

We can now formulate and prove quite easily a general result which takes care of all such examples provided the integrand f is a *positive* function: we need this to ensure that the truncated functions f_n give an increasing sequence. Let us pause at this point to recall that the Lebesgue integral is an *absolute integral* in the sense of Theorem 4.2.2 (iii): if $f \in L^1$ then $|f| \in L^1$ (and $|\int f| \leq \int |f|$).

The reader may have met the famous integral

$$\int_0^\infty \frac{\sin x}{x} dx = \tfrac{1}{2}\pi$$

(most likely in the context of contour integration in complex variable theory). This 'improper' integral is defined in the classical context as

$$\lim_{X \to \infty} \int_0^X \frac{\sin x}{x} dx.$$

(There is no trouble with the integrand $\sin x / x$ at the origin: we assume that the reader is familiar with the fact that this quotient tends to 1 as x tends to 0.) We shall see presently that this limit exists though its evaluation as $\tfrac{1}{2}\pi$ in our context requires more trickery than mathematical insight (see Ex. 5.2.14). But

$$\int_0^\infty \frac{\sin x}{x} dx$$

does *not* exist as a Lebesgue integral because the integral

$$\int_0^\infty \frac{|\sin x|}{x} dx$$

does not exist. We can see this as follows:

$$\int_{(k-1)\pi}^{k\pi} \frac{|\sin x|}{x} dx \geq \frac{1}{k\pi} \int_{(k-1)\pi}^{k\pi} |\sin x| dx = 2/k\pi$$

for $k = 1, 2, \ldots$. Hence

$$\int_0^{k\pi} \frac{|\sin x|}{x} dx \geq \frac{2}{\pi}(1 + 1/2 + 1/3 + \ldots + 1/k)$$

which is unbounded above.

5.1]   MONOTONE CONVERGENCE THEOREM

The convergence process we are using here is the analogue, for integrals, of *absolute convergence*, which was mentioned briefly in §1.1. The following result covers most of the elementary cases that arise in practice. (We say that the sequence $\{I_n\}$ of intervals is *increasing* if $I_n \subset I_{n+1}$ for all $n \geq 1$.)

Proposition 1. *Let $\{I_n\}$ be an increasing sequence of intervals in $\mathbf{R}^k$ whose union is the interval I. If $f \in L^1(I_n)$ for $n \geq 1$ and if the integrals*

$$\int_{I_n} |f|$$

(which exist by Theorem 4.2.2 (iii)) are bounded above, then $f \in L^1(I)$ and

$$\int_I f = \lim \int_{I_n} f.$$

Proof. (i) Consider first the case where f is positive. The 'truncated' functions $f_n = f\chi_{I_n}$ define an increasing sequence $\{f_n\}$ which satisfies the conditions of the Monotone Convergence Theorem and converges to f (everywhere) on I. Thus

$$\int f_n \to \int_I f.$$

(ii) In general we may express f in terms of its positive and negative parts
$$f = f^+ - f^-.$$
Then we may apply part (i) to the positive functions f^+, f^- separately using the truncated functions $f_n{}^+, f_n{}^-$. As their integrals are dominated by $\int |f_n|$ (which are bounded above) we deduce that $f^+, f^- \in L^1(I)$

$$\int f_n{}^+ \to \int_I f^+, \quad \int f_n{}^- \to \int_I f^-,$$

from which $f \in L^1(I)$ and $\quad \int f_n \to \int_I f.$

(Cf. the Proof of Proposition 1.1.2.)

The practical value of Proposition 1 is that it allows us to build up a considerable repertoire of positive integrable functions. But thereafter we shall often check the boundedness of the integrals $\int |f_n|$ by verifying that $\{f_n\}$ is *dominated* by a known positive integrable function g, i.e.
$$|f_n| \leq g \quad (n = 1, 2, \ldots), \tag{1}$$
for then, of course,
$$\int |f_n| \leq \int g \quad (n = 1, 2, \ldots).$$

As a typical example we consider the integral

$$\int_0^X \frac{\sin x}{x} \, dx$$

for large values of X. The integrand is continuous (even at the origin) and so there is no difficulty about

$$\int_0^1 \frac{\sin x}{x} \, dx.$$

We consider

$$\int_1^X \frac{\sin x}{x} \, dx$$

and integrate by parts:

$$\int_1^X \frac{\sin x}{x} \, dx = \left[\frac{-\cos x}{x} \right]_1^X - \int_1^X \frac{\cos x}{x^2} \, dx. \tag{2}$$

(You now see why we had to move away from the origin!) The existence of the integral

$$\int_1^\infty \frac{dx}{x^2}$$

as the limit of

$$\int_1^n \frac{dx}{x^2} = \left[-\frac{1}{x} \right]_1^n = 1 - \frac{1}{n}$$

follows at once from Proposition 1 and then the inequality

$$\left| \frac{\cos x}{x^2} \right| \leq \frac{1}{x^2}$$

for $x \geq 1$ (analogous to (1)) allows us to deduce that

$$\int_1^\infty \frac{\cos x}{x^2} \, dx$$

exists and is the limit of

$$\int_1^{b_n} \frac{\cos x}{x^2} \, dx,$$

where $\{b_n\}$ is any divergent increasing sequence. If we add

$$\int_0^1 \frac{\sin x}{x} \, dx$$

to both sides of (2) we now deduce at once that

$$\lim_{X \to \infty} \int_0^X \frac{\sin x}{x} \, dx$$

exists. The evaluation of this limit as $\frac{1}{2}\pi$ is the subject of Ex. 5.2.14.

5.1]   MONOTONE CONVERGENCE THEOREM

To end this section we prove the formulae for Integration by Parts and Integration by Substitution which generalise the results given in Chapter 3 (Propositions 3.4.1, 2). The earlier results (which make strong assumptions about continuity) are all that are needed in practice and so these proofs are mainly of theoretical interest. But they do show the power of the Monotone Convergence Theorem and also illustrate the generality that can be achieved in this kind of result if we avoid altogether the mention of derivatives.

Proposition 2. *If $f, g \in L^1[a, b]$ and if we define F, G by the equations*

$$F(x) = \int_a^x f + C, \quad G(x) = \int_a^x g + C'$$

for x in $[a, b]$, where C, C' are arbitrary constants, then

$$\int_a^b Fg + \int_a^b fG = [FG]_a^b. \tag{3}$$

Proof. We remark first of all that the constants C, C' do not have any bearing on the truth of equation (3). They make contributions of $C(G(b) - G(a))$, $C'(F(b) - F(a))$, respectively, to each side. We may as well assume that they are both zero. The steps of the proof now follow the steps in the construction of the Lebesgue integral.

(i) If f, g are both constant in $[a, b]$ then we may check the equation at once from first principles. (There is no need to use the earlier result for continuous functions.)

(ii) If f, g are step functions we may express $[a, b]$ as the union of (finitely many) disjoint intervals in each of which f, g are constant and so formula (3) will follow by adding the corresponding formulae which are true for each subinterval.

(iii) If f, g are positive functions in L^{inc} then we may find increasing sequences $\{\phi_n\}$, $\{\psi_n\}$ of step functions which converge almost everywhere to f, g. By considering $\{\phi_n^+\}$, $\{\psi_n^+\}$, if necessary, we may assume that ϕ_n, ψ_n are positive. The formula

$$\int_a^b \Phi_n \psi_n + \int_a^b \phi_n \Psi_n = [\Phi_n \Psi_n]_a^b$$

holds for each n and equation (3) is immediate by the Monotone Convergence Theorem.

(iv) Finally, if $f, g \in L^1$ we may express each of them as the difference almost everywhere of a pair of positive elements from L^{inc} (cf. the proof of Lemma 1 above), and the result follows by linearity.

Proposition 3. Let g be a positive element of $L^1[c,d]$; define
$$G(t) = \int_c^t g + C$$
for t in $[c,d]$ and write $a = G(c)$, $b = G(d)$. If $f \in L^1[a,b]$ then
$$\int_a^b f(x)\,dx = \int_c^d f(G(t))\,g(t)\,dt. \tag{4}$$

Proof. Although the function g may be very wild we recall from §3.5 that G is continuous on $[c,d]$. As g is positive, G is also increasing and assumes each value in $[a,b]$ at least once (see Theorem 3 of the Appendix).

(i) Let $f = \chi_I$ where I is an interval with end points a_1, b_1 contained in $[a,b]$. The set of points t in $[c,d]$ for which $G(t) \in I$ is an interval with end points c_1, d_1, say, where $a_1 = G(c_1)$, $b_1 = G(d_1)$. In this case equation (4) reduces to
$$b_1 - a_1 = G(d_1) - G(c_1).$$

(ii) The result extends at once to step functions by linearity.

(iii) Let $f \in L^{\text{inc}}$ and suppose that the increasing sequence $\{\phi_n\}$ of step functions converges to f outside the null set S. Let T consist of all points t in $[c,d]$ for which $g(t) \neq 0$ and $G(t) \in S$. We shall see below that T is null. Assume this for the moment; then $\{\phi_n(G(t))\,g(t)\}$ is an increasing sequence of functions in $L^1[c,d]$ (Ex. 3.2.6) which converges to $f(G(t))\,g(t)$ for $t \notin T$. Equation (4) now follows by the Monotone Convergence Theorem.

The proof that T is null is similar to the difficult part (iii) of the proof of Fubini's Theorem. As S is null there is an increasing sequence $\{\psi_n\}$ of step functions which diverges on S and whose integrals are bounded (Theorem 3.2.1). Using equation (4) for ψ_n and the Monotone Convergence Theorem we see that the increasing sequence
$$\{\psi_n(G(t))\,g(t)\}$$
must converge almost everywhere on $[c,d]$. But it diverges if $g(t) \neq 0$ and $G(t) \in S$, i.e. if $t \in T$. Hence T is null.

(iv) The extension to L^1 is again immediate from part (iii) by linearity.

Exercises

1. Extend Theorem 1 to the case where $\{f_n\}$ is increasing almost everywhere.

5.1] MONOTONE CONVERGENCE THEOREM

2. If $\chi_S \in L^1$ define the *measure* $m(S)$ of S by the equation

$$m(S) = \int \chi_S.$$

Show that $m(S) = 0$ if and only if S is null.

3. Show that any element of L^1 is equal almost everywhere to the difference of two positive elements of L^{inc}.

4. Use Proposition 1 to prove the existence of

$$\int_0^\infty e^{-x}\,dx \quad \text{and} \quad \int_0^\infty e^{-x^2}\,dx.$$

If the value of the latter is α, say, and if I is the interval (quadrant) in $\mathbf{R}^2$ defined by $x \geq 0$, $y \geq 0$, show that

$$\int_I e^{-(x^2+y^2)}\,d(x,y) = \alpha^2.$$

[See also Ex. 6.4.4.]

5. If
$$f(x) = \frac{1}{x^2}\sin x \quad \text{for} \quad x \geq \pi$$
$$= 0 \quad \text{for} \quad x < \pi$$

show that $f \in L^1(\mathbf{R})$.

6. (i) By considering the intervals $(0, 1)$, $(1, \infty)$ separately show that the integral

$$\Gamma(\alpha) = \int_0^\infty e^{-x} x^{\alpha-1}\,dx$$

exists for $\alpha > 0$.

(ii) Use the inequality of arithmetic and geometric means (Ex. 3.4.5) to show that

$$\left(1 - \frac{x}{n}\right)^n \leq \left(1 - \frac{x}{n+1}\right)^{n+1}$$

provided $|x|/n < 1$.

(iii) Use the logarithmic inequalities of Ex. 3.3.3 and the Monotone Convergence Theorem to show that

$$\int_0^n \left(1 - \frac{x}{n}\right)^n x^{\alpha-1}\,dx \to \Gamma(\alpha).$$

7. If $\{f_n\}$ is a decreasing sequence of positive functions from L^1 for which

$$\int f_n \to 0,$$

show that $f_n \to 0$ almost everywhere.

8. If $a_n \in L^1(\mathbf{R}^k)$ for $n = 1, 2, \ldots$ and if the series

$$\Sigma \int |a_n|$$

is convergent, show that the series Σa_n is absolutely convergent almost everywhere to a function f, where f lies in $L^1(\mathbf{R}^k)$ and where the series

$$\Sigma \int a_n$$

is absolutely convergent to $\int f$. (Cf. the proof of Proposition 1.1.2.)

9. Prove that
$$\int_0^\infty \frac{e^{-x}}{1-e^{-x}} x^{\alpha-1} dx = \Gamma(\alpha) \sum_{n=1}^\infty \frac{1}{n^\alpha}$$

for $\alpha > 1$. (See Ex. 6 for the definition of $\Gamma(\alpha)$.)

Improper Riemann Integrals

There is a whole theory of *improper* Riemann integrals which we shall illustrate by giving two particular cases.

(i) If
$$(R) \int_a^b f(x) \, dx$$

exists for all a, b in $\mathbf{R}$ and if these integrals tend to a limit l as $a \to -\infty$, $b \to \infty$ (independently) then we write

$$(R) \int_{-\infty}^\infty f(x) \, dx = l.$$

Show that
$$(R) \int_{-\infty}^\infty e^{-|x|} \, dx = 2.$$

(ii) If
$$(R) \int_c^b f(x) \, dx$$

exists for all c in $(a, b]$ and if these integrals tend to a limit l as $c \searrow a$, then we write

$$(R) \int_a^b f(x) \, dx = l.$$

Show that
$$(R) \int_0^1 x^{-\frac{1}{2}} \, dx = 2.$$

5.2 The Dominated Convergence Theorem

The main result of this section is even more famous than the Monotone Convergence Theorem, as it was the cornerstone of the original treatment of integration by Lebesgue [12]. The Monotone Convergence

Theorem was published a few years later by Beppo Levi [15]. At a first glance (see Theorem 1) the Dominated Convergence Theorem appears to be considerably more general than the Monotone Convergence Theorem as it refers to the general notion of convergence rather than the much simpler notion of convergence for monotone sequences. Nevertheless we shall see that Theorem 1 may be deduced with very little trouble from the Monotone Convergence Theorem once we show how to discuss a quite general convergent sequence in terms of monotone sequences.

The reader will appreciate the fact that we have used the idea of monotone convergence or, at worst, absolute convergence, throughout the book – so far! The Axiom of Completeness and the construction of the Lebesgue integral were entirely framed in terms of this intuitively simple concept. Suppose now, however, that we are given a sequence $\{s_n\}$ of real numbers and all we know is that its terms are *bounded*, say
$$-K \leqslant s_n \leqslant K$$
for all n. We may construct two monotone sequences as follows. Let
$$u_n = \sup\{s_n, s_{n+1}, \ldots\}.$$
This supremum exists by Theorem 1.2.2. Also
$$u_n \geqslant u_{n+1}$$
for all n, so that the sequence $\{u_n\}$ is *decreasing*. In exactly the same way we define
$$l_n = \inf\{s_n, s_{n+1}, \ldots\}$$
and see at once that
$$l_n \leqslant l_{n+1}$$
for all n, i.e. the sequence $\{l_n\}$ is *increasing*.

From these definitions it is clear that
$$l_n \leqslant u_n$$
for all n. But if $m \leqslant n$ we see that
$$l_m \leqslant l_n \leqslant u_n \leqslant u_m$$
and so
$$l_m \leqslant u_n$$
for all m, n. It is natural to call $\{l_n\}$ the *lower sequence* and $\{u_n\}$ the *upper sequence* for $\{s_n\}$. Moreover, these bounded monotone sequences converge to limits l, u, say, called the *lower* and *upper limits* for $\{s_n\}$ which satisfy $l \leqslant u$. (Apply Proposition 1.1.1 (v) to the inequality $l_n \leqslant u_n$.) A self explanatory notation that is often used is
$$l = \liminf s_n, \quad u = \limsup s_n.$$

There is much more that can be said about these fruitful ideas but our modest purpose in introducing them is to show how results on general convergent sequences may be deduced from results on monotone sequences. First of all we need a very elementary fact which shows that we can apply the above ideas to an arbitrary convergent sequence.

Lemma 1. *A convergent sequence of real numbers is bounded.*

Proof. Let $\{s_n\}$ converge to s. Take $\epsilon = 1$ in the definition of convergence; then there exists an integer N such that
$$|s_n - s| < 1$$
for all $n \geqslant N$, i.e.
$$s - 1 < s_n < s + 1$$
for all $n \geqslant N$. Now clearly $\{s_n\}$ is bounded above by the largest of the numbers $s_1, s_2, \ldots, s_{N-1}, s+1$ and bounded below by the smallest of the numbers $s_1, s_2, \ldots, s_{N-1}, s-1$.

The key result for our purpose is:

Proposition 1. *The bounded sequence $\{s_n\}$ of real numbers converges to the limit s if and only if*
$$\liminf s_n = \limsup s_n = s.$$

Proof. (i) We are given a bounded sequence $\{s_n\}$ and suppose that $\liminf s_n = \limsup s_n = s$, say. Then we have the sequence $\{s_n\}$ 'squeezed' between the sequences $\{l_n\}, \{u_n\}$, i.e.
$$l_n \leqslant s_n \leqslant u_n$$
for all n, where $\{l_n\}$ and $\{u_n\}$ both converge to the same limit s. Thus $\{s_n\}$ converges to s.

(ii) Suppose that $\{s_n\}$ converges to s. Given $\epsilon > 0$; there is an integer N such that
$$|s_n - s| < \epsilon$$
whenever $n \geqslant N$. Thus $s - \epsilon$ and $s + \epsilon$ are lower and upper bounds for $\{s_N, s_{N+1}, \ldots\}$. By the definition of the lower and upper sequences
$$s - \epsilon \leqslant l_N \leqslant u_N \leqslant s + \epsilon.$$
This gives
$$s - \epsilon \leqslant \liminf s_n \leqslant \limsup s_n \leqslant s + \epsilon.$$
But now $\liminf s_n$ and $\limsup s_n$ are fixed real numbers and ϵ is arbitrarily small, so we have another 'squeezing' argument which shows that
$$s \leqslant \liminf s_n \leqslant \limsup s_n \leqslant s,$$
i.e.
$$\liminf s_n = \limsup s_n = s.$$

We are now in a position to prove our main result.

DOMINATED CONVERGENCE THEOREM

Theorem 1 (Lebesgue's Theorem of Dominated Convergence).
Suppose that $\{f_n\}$ is a sequence of integrable functions which converges almost everywhere to a function f and which is 'dominated' by an integrable function g in the sense that

$$|f_n| \leq g$$

for all n. Then f is integrable and

$$\int f_n \to \int f.$$

Proof. Let x be a point of $\mathbf{R}^k$ for which $f_n(x) \to f(x)$. We may define

$$l_n(x) = \inf\{f_n(x), f_{n+1}(x), \ldots\},$$
$$u_n(x) = \sup\{f_n(x), f_{n+1}(x), \ldots\}$$

as above, and this can be done for almost all x. As usual we define

$$l_n(x) = 0, \quad u_n(x) = 0$$

for any 'bad' points x for which $\{f_n(x)\}$ fails to converge to $f(x)$.

We now have two *monotone* sequences of functions $\{l_n\}$, $\{u_n\}$ which both converge to f almost everywhere and which satisfy

$$l_n \leq f_n \leq u_n$$

for all n almost everywhere. Let us assume for the moment that l_n and u_n are integrable (for all n), then the Monotone Convergence Theorem shows that f is integrable and

$$\int l_n \to \int f, \quad \int u_n \to \int f.$$

But
$$\int l_n \leq \int f_n \leq \int u_n$$

and our familiar squeezing argument shows that

$$\int f_n \to \int f.$$

All that remains to prove is that l_n and u_n are integrable for all n. If f_1, f_2 are both integrable then

$$\min\{f_1, f_2\} = \tfrac{1}{2}(f_1 + f_2) - \tfrac{1}{2}|f_1 - f_2|,$$
$$\max\{f_1, f_2\} = \tfrac{1}{2}(f_1 + f_2) + \tfrac{1}{2}|f_1 - f_2|$$

are also integrable. This extends at once by induction to a finite number of functions. Thus

$$\min\{f_n, f_{n+1}, \ldots, f_{n+k}\} = l_{nk},$$
$$\max\{f_n, f_{n+1}, \ldots, f_{n+k}\} = u_{nk}$$

are integrable. If we keep n fixed and allow $k = 1, 2, \ldots$ to vary we obtain two monotone sequences whose limits almost everywhere are l_n and u_n (see Theorem 1.2.1). Moreover the domination by the integrable function g ensures that

$$\left|\int l_{nk}\right| \leq \int g, \quad \left|\int u_{nk}\right| \leq \int g$$

and so the Monotone Convergence Theorem shows that l_n and u_n are integrable. This completes the proof. (Note that the dominating function g was only used at the end of the proof to show that l_n and u_n are integrable.)

In many practical situations one is interested in finding integrals on a bounded interval and the following special case of Theorem 1 is extremely useful.

Corollary (Lebesgue's Theorem of Bounded Convergence). *Suppose that f_n is integrable on the bounded interval I for all n and that the sequence $\{f_n\}$ converges almost everywhere on I to a function f. If there is a real number K such that*

$$|f_n(x)| \leq K$$

for all n, and all x in I, then f is integrable on I and

$$\int_I f_n \to \int_I f.$$

Proof. Let $g = K\chi_I$. Then we may apply Theorem 1 to the sequence $\{f_n \chi_I\}$.

There is a third very well known result in which we are given a convergent sequence $\{f_n\}$ of integrable functions, but no dominating function g, and we are able to deduce correspondingly less about the integral of the limit function.

Theorem 2 (Fatou's Lemma). *Suppose that $\{f_n\}$ is a sequence of positive integrable functions which converges almost everywhere to a function f and that there is a real number K such that*

$$\int f_n \leq K$$

5.2] DOMINATED CONVERGENCE THEOREM

for all n. Then f is integrable and

$$\int f \leq K.$$

Note that we have assumed that the functions f_n are *positive*. This condition could easily be missed in a casual reading of the theorem; it ensures that the integrals $\int f_n$ are bounded below (by zero).

Proof. This is practically the same as the proof of Theorem 1, but we use only the lower sequence $\{l_n\}$. Just as before, the functions

$$l_{nk} = \min\{f_n, f_{n+1}, \ldots, f_{n+k}\}$$

are integrable, and for n fixed and $k = 1, 2, \ldots$ give a decreasing sequence which converges almost everywhere to l_n. Furthermore, our given inequality for the integrals ensures that

$$0 \leq \int l_{nk} \leq K$$

and so the Monotone Convergence Theorem shows that l_n is integrable and

$$0 \leq \int l_n \leq K.$$

But the increasing sequence $\{l_n\}$ converges to f almost everywhere, so a final application of the Monotone Convergence Theorem shows that f is integrable and

$$\int f \leq K.$$

The final inequality in Fatou's Lemma is often given as

$$\int f \leq \liminf \int f_n.$$

For the proof and an example where the strict inequality holds see Ex. 5.

As a good example of the use of the Dominated Convergence Theorem we mention the integral form of the Gamma Function

$$\Gamma(\alpha) = \int_0^\infty e^{-x} x^{\alpha-1} dx \quad (\alpha > 0).$$

(The existence of this integral is shown in Ex. 5.1.6.) There is another expression for $\Gamma(\alpha)$ as the limit of the product

$$P_n(\alpha) = \frac{n!\, n^\alpha}{\alpha(\alpha+1)\ldots(\alpha+n)} \quad (\alpha > 0)$$

as $n \to \infty$. It is easily verified (Ex. 6) that
$$\int_0^1 (1-t)^n t^{\alpha-1} dt = \frac{n!}{\alpha(\alpha+1)\ldots(\alpha+n)}$$
and so, by the substitution $x = tn$,
$$P_n(\alpha) = \int_0^n \left(1-\frac{x}{n}\right)^n x^{\alpha-1} dx.$$
The equivalence of the two definitions of $\Gamma(\alpha)$ now depends on showing that
$$\int_0^n \left(1-\frac{x}{n}\right)^n x^{\alpha-1} dx \to \int_0^\infty e^{-x} x^{\alpha-1} dx.$$
But this is immediate by the Dominated Convergence Theorem as
$$\left(1-\frac{x}{n}\right)^n \leq e^{-x} \quad \text{for} \quad 0 \leq x \leq n$$
and
$$\left(1-\frac{x}{n}\right)^n \to e^{-x}.$$

(See Ex. 5.1.6 for details.) The student may like to compare this proof with the classical one given in, say [28] or [1].

We promised earlier to give a more general form of the Fundamental Theorem of the Calculus (cf. Theorem 3.4.2). This is now comparatively simple by means of the Bounded Convergence Theorem.

Theorem 3. *If the function F has bounded derivative f on the closed interval $[a, b]$, then f is integrable on $[a, b]$ and*
$$\int_a^b f(x) dx = F(b) - F(a).$$

The above condition means that $F'(x) = f(x)$ exists for *all* x in $[a, b]$, where the derivatives at a, b, respectively, are interpreted as right and left hand derivatives, and
$$|f(x)| \leq K$$
for all x in $[a, b]$.

Proof. If $a = x_0 < x_1 < x_2 < \ldots < x_r = b$ let ϕ be the step function which takes the constant value (average gradient)
$$\frac{F(x_i) - F(x_{i-1})}{x_i - x_{i-1}}$$
on the interval $[x_{i-1}, x_i)$ for $i = 1, 2, \ldots, r$, and 0 elsewhere. Then
$$\int \phi = \sum_{i=1}^r \{F(x_i) - F(x_{i-1})\} = F(b) - F(a).$$

5.2] DOMINATED CONVERGENCE THEOREM

Moreover, the Mean Value Theorem shows that

$$\frac{F(x_i) - F(x_{i-1})}{x_i - x_{i-1}} = f(c_i)$$

for some point c_i in (x_{i-1}, x_i) and so

$$|\phi| \leq K.$$

To each subdivision of $[a, b)$ there is such a step function ϕ. If we subdivide $[a, b)$ by successive halving we obtain a sequence $\{\phi_n\}$ of step functions vanishing outside $[a, b)$ for which

$$\int \phi_n = F(b) - F(a)$$

and

$$|\phi_n| \leq K.$$

Let p be a point of $[a, b)$. Given $\epsilon > 0$, there is a $\delta > 0$ such that

$$f(p) - \epsilon < \frac{F(x) - F(p)}{x - p} < f(p) + \epsilon$$

for $0 < |x - p| < \delta$ and $x \in [a, b]$. Thus

$$(f(p) - \epsilon)(x - p) < F(x) - F(p) < (f(p) + \epsilon)(x - p)$$

for $0 < x - p < \delta$ and

$$(f(p) - \epsilon)(p - x) < F(p) - F(x) < (f(p) + \epsilon)(p - x)$$

for $0 < p - x < \delta$ ($x \in [a, b]$). From this it follows that if the interval $[x_{i-1}, x_i)$ contains p and if $x_i - x_{i-1} < \delta$ then

$$(f(p) - \epsilon)(x_i - x_{i-1}) < F(x_i) - F(x_{i-1}) < (f(p) + \epsilon)(x_i - x_{i-1})$$

and

$$f(p) - \epsilon < \frac{F(x_i) - F(x_{i-1})}{x_i - x_{i-1}} < f(p) + \epsilon.$$

Now it is clear that $\phi_n(x) \to f(x)$ for all x in $[a, b)$ and an application of the Theorem of Bounded Convergence gives

$$\int_a^b f(x) \, dx = F(b) - F(a).$$

We give just one more application of the Dominated Convergence Theorem which is of great practical value (not least because it will point the way in the next chapter to the important concept of a measurable function).

Proposition 2. *Let $f: \mathbf{R}^k \to \mathbf{R}$ be the limit almost everywhere of a sequence of integrable functions, and suppose that*

$$|f| \leq g$$

for some positive integrable function g. Then f is integrable.

Notation: If a, b, c are any three real numbers (not necessarily distinct) then
$$\mathrm{mid}\{a, b, c\}$$
denotes the unique number among a, b, c that is between the other two. (See Ex. 8.)

Proof. Let $\{f_n\}$ be a sequence of integrable functions which converges almost everywhere to f, and define the function h_n as follows:

$$h_n = \mathrm{mid}\{-g, f_n, g\},$$

i.e.
$$h_n(x) = \mathrm{mid}\{-g(x), f_n(x), g(x)\}$$

for all x. This is just another way of saying that h_n is the function obtained by 'truncating' f_n above by g and below by $-g$ (see Fig. 39).

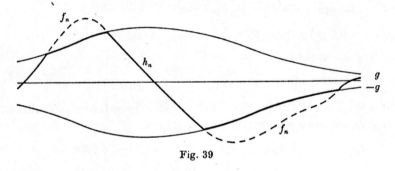

Fig. 39

We may also write $\quad h_n = \max\{-g, \min\{f_n, g\}\}$

and so h_n is integrable. Moreover

$$|f| \leq g$$

so that $h_n(x)$ is at least as good an approximation to $f(x)$ as $f_n(x)$. Thus
$$h_n \to f$$
almost everywhere, and $\quad |h_n| \leq g.$

The integrability of f now follows by Theorem 1.

As an example of the use of Proposition 2 we can prove a famous result from the theory of Fourier Series.

The Riemann–Lebesgue Lemma. *If $f \in L^1(\mathbf{R})$ then the integrals*

$$\int_{-\infty}^{\infty} f(x) \cos kx \, dx,$$

$$\int_{-\infty}^{\infty} f(x) \sin kx \, dx$$

both exist and converge to 0 as $k \to \infty$.

Proof. We consider only the cosine integral as the other case is almost identical. There are the customary four steps.

(i) If $f = \chi_I$ where I is an interval with end points a, b then

$$\int_a^b \cos kx \, dx = \frac{1}{k}(\sin kb - \sin ka)$$

which converges to 0 as $k \to \infty$ because $|\sin x| \leqslant 1$ for all x.

(ii) The lemma extends to step functions by linearity.

(iii) If $f \in L^{\text{inc}}$ then there is an increasing sequence $\{\phi_n\}$ of step functions which converges to f almost everywhere and

$$\int \phi_n \to \int f.$$

Now $$\phi_n(x) \cos kx \to f(x) \cos kx$$

for almost all x and $\quad |f(x) \cos kx| \leqslant |f(x)|$

for all x, so Proposition 2 shows that

$$\int_{-\infty}^{\infty} f(x) \cos kx \, dx$$

exists.

Given $\epsilon > 0$; there is an integer N such that

$$0 \leqslant \int (f - \phi_N) < \epsilon.$$

Having chosen N we can then find a real number K (depending on N and ultimately on ϵ) such that

$$\left| \int \phi_N(x) \cos kx \, dx \right| < \epsilon$$

for $k \geq K$. Finally

$$\left|\int f(x)\cos kx\,dx\right| \leq \left|\int (f(x)-\phi_N(x))\cos kx\,dx\right| + \left|\int \phi_N(x)\cos kx\,dx\right|$$

$$< \int (f-\phi_N) + \epsilon$$

$$< 2\epsilon$$

for $k \geq K$. This proves the lemma if $f \in L^{\text{inc}}$.

(iv) The extension to L^1 is immediate by linearity.

Exercises

1. *(Cauchy's General Principle of Convergence.)* A sequence $\{s_n\}$ of real numbers satisfies the following condition: given $\epsilon > 0$; there is an integer N such that $|s_m - s_n| < \epsilon$ whenever $m, n \geq N$. Show that $\{s_n\}$ is bounded and

$$s_N - \epsilon \leq \liminf s_n \leq \limsup s_n \leq s_N + \epsilon.$$

Hence show that $\{s_n\}$ is convergent. (Cf. Lemma 1 and Proposition 1.)

2. Let
$f_n(x) = nx/(1+n^2x^2)$ for x in $[0,1]$,
$g_n(x) = n^{\frac{3}{2}}x/(1+n^2x^2)$ for x in $[0,1]$,
$h_n(x) = n^{\frac{3}{2}}x^{\frac{3}{4}}/(1+n^2x^2)$ for x in $[0,1]$.

Show that the Theorem of Bounded Convergence applies to $\{f_n\}$ and the Theorem of Dominated Convergence applies to $\{g_n\}$.

3. Let
$$f(x) = x^{-\frac{1}{2}} \text{ for } x \text{ in } (0,4],$$
$$= 0 \text{ otherwise.}$$

What functions are obtained if we truncate f above by $n\chi_{(0,4]}$ ($n = 1, 2, \ldots$)? Use these functions to evaluate

$$\int_0^4 x^{-\frac{1}{2}}\,dx.$$

4. Give an example of a bounded sequence of integrable functions $\{f_n\}$ that converges almost everywhere to a non-integrable function f.

5. Extend the proof of Fatou's Lemma to show that

$$\int l_n \leq \liminf \int f_n$$

and hence show that

$$\int f \leq \liminf \int f_n.$$

Give an example for which Fatou's Lemma applies with the strict inequality.

6. Let
$$I_{n,\alpha} = \int_0^1 (1-t)^n t^{\alpha-1} dt$$
for $\alpha > 0$, $n = 0, 1, 2, \ldots$. Show that
$$I_{n,\alpha} = \frac{n}{\alpha} I_{n-1,\alpha+1} \quad (n \geq 1)$$
and hence find an expression for $I_{n,\alpha}$.

7. If $F(x) = x^2 \sin(1/x^2)$ for $x > 0$ show that $F'(x) = f(x)$ exists for all $x > 0$ but that f is not integrable on $(0, 1)$.

8. Show that
$$\text{mid}\{a, b, c\} = \max\{\min\{b, c\}, \min\{c, a\}, \min\{a, b\}\}$$
$$= \min\{\max\{b, c\}, \max\{c, a\}, \max\{a, b\}\}.$$
Hence show that if f, g, h are integrable then so is mid $\{f, g, h\}$.

9. Use Proposition 2 to show that if f, g are integrable and f is bounded, then fg is integrable. (See Ex. 3.2.6.)

Give an example to show that the assumption about the boundedness of f cannot be dropped.

10. *Alternative proof of Theorem 3.* Redefine F, if necessary, so that $F(x) = F(b)$ for $x > b$, and define
$$f_n(x) = n\left\{F\left(x + \frac{1}{n}\right) - F(x)\right\}$$
for x in $[a, b]$. Use the Theorem of Bounded Convergence to show that
$$\int_a^b f_n(x)\, dx \to \int_a^b f(x)\, dx$$
and hence prove Theorem 3.

11. Recall Ex. 3.5.1. We saw there that if $f \in L^1[a, b]$ and $F(x) = \int_a^x f + C$ for x in $[a, b]$, then the *total variation* T of F on $[a, b]$ satisfies
$$T \leq \int_a^b |f|.$$
Let $\{\phi_n\}$ be a sequence of step functions that converges to f almost everywhere on $[a, b]$ and define
$$\epsilon_n(x) = +1 \quad \text{if} \quad \phi_n(x) > 0,$$
$$= 0 \quad \text{if} \quad \phi_n(x) = 0,$$
$$= -1 \quad \text{if} \quad \phi_n(x) < 0.$$

By considering the integrals $\int_a^b \epsilon_n f$

show that $T = \int_a^b |f|.$

Differentiation under the integral sign

In the following three exercises we use the classical notation for functions.

12. Suppose that $f(x,t)$ is integrable as a function of x for each value of t and differentiable as a function of t for each value of x, and that

$$\left|\frac{\partial}{\partial t} f(x,t)\right| \leq g(x)$$

for all x,t where $g(x)$ is an integrable function of x. Show that $\frac{\partial}{\partial t} f(x,t)$ is integrable as a function of x for each t and

$$\frac{d}{dt} \int f(x,t)\,dx = \int \frac{\partial}{\partial t} f(x,t)\,dx.$$

[Hint. First prove the *Lemma*. If $f(h_n) \to l$ for every sequence $\{h_n\}$ of non-zero real numbers converging to 0, then

$$f(h) \to l \quad \text{as} \quad h \to 0.$$

Then apply the Mean Value Theorem to

$$\frac{1}{h}(f(x,t+h) - f(x,t)).]$$

13. Adapt the result of Ex. 12 to the case where t is restricted to a bounded closed interval $[c,d]$ and the derivatives at c,d are suitably one-sided.

14. Let
$$F(t) = \int_0^\infty e^{-tx} \frac{\sin x}{x}\,dx \quad (t > 0).$$

(i) Use Ex. 13 to show that
$$F'(t) = -1/(1+t^2) \quad (t > 0).$$

(ii) Show that $\quad F(t) = C - \tan^{-1} t \quad (t > 0),$

where C is a constant.

(iii) By considering the sequence $\{F(n)\}$ show that $C = \tfrac{1}{2}\pi$.

(iv) Show that
$$\int_0^X \frac{\sin x}{x}\,dx \to \tfrac{1}{2}\pi \quad \text{as} \quad X \to \infty.$$

(See Ex. 6.1.10 for a less complicated proof.)

6

MEASURABLE FUNCTIONS AND LEBESGUE MEASURE

Some idea of Lebesgue measure has already been given in Chapter 4, in order to illustrate Fubini's Theorem. The powerful convergence theorems of Chapter 5 now give immediate proofs of the fundamental properties of this measure (Theorem 6.2.1). We could therefore go directly to a full discussion of Lebesgue measure, but we prefer to set aside one more section on the very fruitful idea of *measurable function* and to formulate the general definition of measure in terms of measurable functions. (An important reason for doing this will emerge when we discuss the Daniell integral in the second volume.) One deficiency of our definition of measure is that it depends on the special frame of reference for $\mathbf{R}^k$ and the fact that intervals all have their sides parallel to the 'coordinate axes'. In §6.3 we show that Lebesgue measure is invariant under all Euclidean transformations; in so doing we find that any linear mapping L multiplies the measure of any set by a constant ($|\det L|$) and transforms integrals in a simple way (Theorem 6.3.2). In the last section this result is extended to give the famous *Jacobian* formula for the transformation of integrals in $\mathbf{R}^k$.

6.1 Measurable functions

We have already stated that the linear space L^1 of integrable functions is 'very large' (see the closing paragraph of §3.2) but even so, it does not contain all the 'very good' functions of analysis. For example, it does not contain all of the continuous functions or even the non-zero constant functions. Moreover, the product of two integrable functions may not be integrable, witness the function f defined by

$$f(x) = x^{-\frac{1}{2}} \quad \text{for} \quad 0 < x \leqslant 1.$$
$$= 0 \quad \text{elsewhere,}$$

which is integrable although f^2 is not.

Roughly speaking, some otherwise satisfactory functions fail to be integrable, either because they are unbounded in too drastic a way or because they are not small enough outside a bounded interval I. A

solution to this problem is suggested by the method of *truncating* introduced in Proposition 5.2.2 at the end of the previous chapter. We shall say that a function $f: \mathbf{R}^k \to \mathbf{R}$ is *measurable* if the truncated function
$$\mathrm{mid}\{-g, f, g\}$$
is integrable for every positive integrable function g in $L^1(\mathbf{R}^k)$ (see Fig. 40).

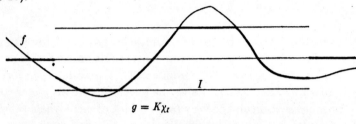

$g = K\chi_I$

Fig. 40

See also the Fig. 39 on p. 114.

On the face of it, this definition would seem to impose an intolerable burden on anyone wanting to check whether or not a given function f is measurable. We reduce the apparent burden by noting that the positive integrable functions g in the definition may be replaced by:

(i) *positive step functions* ϕ;

or

(ii) *step functions $K\chi_I$ where K is positive and I is a bounded interval.*

Let us assume that $\mathrm{mid}\{-\phi, f, \phi\}$ is integrable for any positive step function ϕ. If g is a positive integrable function then we may find a sequence $\{\phi_n\}$ of positive step functions converging almost everywhere to g (Ex. 3.2.5). Thus
$$\mathrm{mid}\{-\phi_n, f, \phi_n\} \to \mathrm{mid}\{-g, f, g\}$$
almost everywhere (Ex. 1.1.2 and Ex. 5.2.8). But
$$|\mathrm{mid}\{-g, f, g\}| \leqslant g$$
and so $\mathrm{mid}\{-g, f, g\}$ is integrable by Proposition 5.2.2. This justifies the first reduction.

The other reduction is much easier. For any positive step function ϕ we can find a step function $K\chi_I$ satisfying $K\chi_I \geqslant \phi$. If we know that the truncated function
$$\mathrm{mid}\{-K\chi_I, f, K\chi_I\}$$
is integrable, then we may truncate again by ϕ to obtain
$$\mathrm{mid}\{-\phi, f, \phi\}$$
which must also be integrable (Ex. 5.2.8).

6.1] MEASURABLE FUNCTIONS

The following result is a major step in showing that the measurable functions include all the functions $f: \mathbf{R}^k \to \mathbf{R}$ that we ever normally encounter in the study of analysis.

Theorem 1. (i) *All continuous and all integrable functions are measurable.*

(ii) *If f, g are measurable then so are $|f|$, f^+, f^-, $af+bg$ $(a,b \in \mathbf{R})$, $\max\{f,g\}$ and $\min\{f,g\}$.*

(iii) *If $f_n \to f$ almost everywhere and f_n is measurable for $n = 1, 2, \ldots$ then f is measurable.*

Proof. (i) If f is continuous, Theorem 4.2.4', Corollary shows that $f\chi_I$ is integrable and so also is mid$\{-K\chi_I, f, K\chi_I\}$.

If f is integrable then any function mid$\{-\phi, f, \phi\}$ is integrable.

(ii) Let ϕ be a positive step function. Then
$$\mathrm{mid}\{-\phi, |f|, \phi\} = |\mathrm{mid}\{-\phi, f, \phi\}|$$
which is integrable and so $|f|$ is measurable.

Let f_n, g_n be the integrable functions defined by
$$f_n = \mathrm{mid}\{-n\phi, f, n\phi\},$$
$$g_n = \mathrm{mid}\{-n\phi, g, n\phi\}.$$
If $\phi(x) = 0$, $f_n(x) = g_n(x) = 0$, and if $\phi(x) > 0$, $f_n(x) = f(x)$, $g_n(x) = g(x)$ for $n \geq N$ (where N depends on x). Thus
$$\mathrm{mid}\{-\phi, af_n + bg_n, \phi\} \to \mathrm{mid}\{-\phi, af+bg, \phi\}.$$
The functions on the left hand side are integrable and are dominated by ϕ. The Dominated Convergence Theorem therefore shows that mid$\{-\phi, af+bg, \phi\}$ is integrable and so $af+bg$ is measurable.

The measurability of f^+, f^-, $\max\{f,g\}$, $\min\{f,g\}$ now follows from the simple relations
$$f^+ = \tfrac{1}{2}(|f|+f),$$
$$f^- = \tfrac{1}{2}(|f|-f),$$
$$\max\{f,g\} = \tfrac{1}{2}(f+g) + \tfrac{1}{2}|f-g|,$$
$$\min\{f,g\} = \tfrac{1}{2}(f+g) - \tfrac{1}{2}|f-g|.$$

(iii) The functions mid$\{-\phi, f_n, \phi\}$ are integrable for $n = 1, 2, \ldots$, are dominated by ϕ, and mid$\{-\phi, f_n, \phi\} \to \mathrm{mid}\{-\phi, f, \phi\}$ almost everywhere. Thus mid$\{-\phi, f, \phi\}$ is integrable and so f is measurable.

As a supplement to Theorem 1 we also note here that *the product of any two measurable functions is measurable*: this will be proved in Ex. 11 and again in Proposition 6.2.3, Corollary.

In order to show as vividly as possible the scope of this class of measurable functions we mention a hierarchy of functions that was studied at the turn of the century by Baire [3]. The continuous functions (on $\mathbf{R}^k$) form the *Baire class* 0; the functions which are limits (everywhere) of convergent sequences of continuous functions form the *Baire class* 1; the functions which are limits (everywhere) of convergent sequences of functions of Baire class 1 form the *Baire class* 2; and so on. This 'and so on' is not as innocent as it might appear. There is no difficulty in showing that the Baire class α is contained in the Baire class $(\alpha+1)$ for each α (simply consider sequences in which all the terms are equal to a function of Baire class α). But, in fact, there is a *strictly* increasing chain which goes on transfinitely as far as the first non-countable ordinal number Ω. This last statement is far too hard to establish in this book (see [8], Ch. 11). Even the proof that the Baire class 2 is strictly greater than the Baire class 1 is far from trivial (cf. Ex. 9). Despite the complexity of this construction it follows from Theorem 1 that all these Baire classes are contained in our space of measurable functions. It almost goes without saying that the Baire classes incorporate most of the functions studied in elementary analysis. Our point in introducing them was not to frighten the reader but rather to inspire his confidence that the class of measurable functions which we have constructed is so comprehensive that he need never wander out of it!

In practice the verification of the measurability of a given function f hardly ever involves our definition directly, but is usually quite obvious from Theorem 1. In fact the criterion, *par excellence*, for integrability is now the following result, which would have appeared ridiculous if placed immediately after our definition of measurability without the power of Theorem 1.

Proposition 1. *If f is measurable and if $|f| \leq g$ where g is integrable, then f is integrable.*

Proof. $\operatorname{mid}\{-g, f, g\} = f$!

Corollary. *If f is measurable and $|f|$ is integrable, then f is integrable.*

For example, if $f \in L^1(R)$ the existence of the integrals

$$\int_{-\infty}^{\infty} f(x) \cos kx \, dx,$$

$$\int_{-\infty}^{\infty} f(x) \sin kx \, dx$$

follows from Proposition 1 as the integrands are measurable (Ex. 11) and are dominated by $|f|$. (See the Riemann-Lebesgue Lemma in §5.2.)

There is another famous result which can now be stated in terms of measurable functions and which has many practical applications in the evaluation of integrals on $\mathbf{R}^2$. This is a partial converse of Fubini's Theorem.

Theorem 2 (*Tonelli's Theorem*). *If $f: \mathbf{R}^2 \to \mathbf{R}$ is measurable and if one of the repeated integrals*

$$\int \left(\int |f(x,y)|\, dx \right) dy, \quad \int \left(\int |f(x,y)|\, dy \right) dx$$

exists, then f is integrable and hence the repeated integrals

$$\int \left(\int f(x,y)\, dx \right) dy, \quad \int \left(\int f(x,y)\, dy \right) dx$$

both exist and are equal.

Proof. Define the closed interval I_n in $\mathbf{R}^2$ by the inequalities

$$|x| \leq n, \quad |y| \leq n.$$

The truncated functions

$$h_n = \mathrm{mid}\{-n\chi_{I_n}, |f|, n\chi_{I_n}\} = \min\{|f|, n\chi_{I_n}\}$$

are integrable (on $\mathbf{R}^2$) and satisfy

$$\int h_n = \int \left(\int h_n(x,y)\, dx \right) dy = \int \left(\int h_n(x,y)\, dy \right) dx$$

by Fubini's Theorem. But the increasing sequence $\{h_n\}$ converges to $|f|$ everywhere and $\int h_n$ is bounded above by one of the given repeated integrals, so the Monotone Convergence Theorem proves that $|f|$ is integrable. The above Corollary to Proposition 1 now shows that f is integrable and the rest follows from the Corollary to Fubini's Theorem.

Exercises

1. Let f be a measurable function on $\mathbf{R}^k$, $a \in \mathbf{R}^k$, $t \in \mathbf{R}$; define

$$g(x) = f(x+a), \quad h(x) = f(tx)$$

for all x in $\mathbf{R}^k$. Show that g, h are measurable.

2. Let $f, g, h: \mathbf{R}^k \to \mathbf{R}$. If f is measurable and g, h are integrable, show that $\mathrm{mid}\{f, g, h\}$ is integrable.

3. If f is a bounded integrable function on $\mathbf{R}^k$ show that f^2 is integrable.

4. If f^2 is measurable show that $|f|$ is measurable. Give an example where f^2 is integrable but $|f|$ is not integrable.

5. Let f, g be measurable functions on $\mathbf{R}^k$. Show that f, g are integrable if and only if $\sqrt{(f^2+g^2)}$ is integrable.

6. Let f, g be measurable functions on $\mathbf{R}^k$. If f^2, g^2 are integrable, show that fg is integrable. By considering the function $(af+bg)^2$ for all a, b in $\mathbf{R}$ show that
$$\left(\int fg\right)^2 \leq \int f^2 \int g^2.$$

7. Let $\{f_n\}$ be a bounded sequence of measurable functions on $\mathbf{R}^k$. Show that $\liminf f_n$, $\limsup f_n$ are measurable functions.

8. Show that the step functions on $\mathbf{R}^k$ all belong to the Baire class 1.

9. Show that the characteristic function of the rationals in $\mathbf{R}^1$ belongs to the Baire class 2. (By a 'category' argument Baire showed that this function does not belong to the Baire class 1.)

10. Use Tonelli's Theorem to show that
$$\int_0^X \frac{\sin x}{x} dx = \int_0^\infty \left(\int_0^X e^{-xy} \sin x \, dx\right) dy$$
for $X > 0$. If X tends to ∞ (through a sequence of values) show that
$$\int_0^X \frac{\sin x}{x} dx \to \tfrac{1}{2}\pi.$$
(Cf. Ex. 5.2.14.)

11. (i) Let the rationals be arranged as the terms of a sequence $\{r_n\}$. Show that
$$\max\{2r_k t - r_k^2 : 1 \leq k \leq n\} \to t^2$$
as $n \to \infty$.

(ii) Let f be a measurable function; deduce from (i) that f^2 is measurable.

(iii) Let f, g be measurable functions. Use the identity
$$4fg = (f+g)^2 - (f-g)^2$$
to deduce that fg is measurable.

12. Given f in $L^1[a, b]$; use Exercise 11 and the above Proposition to show that
$$I_n = \int_a^b f(x) (\sin x)^n \, dx$$
exists for each positive integer n. Show also that $I_n \to 0$ as $n \to \infty$.

6.2 Lebesgue measure on $\mathbf{R}^k$

Let S be a set of points in $\mathbf{R}^k$ with characteristic function $\chi_S : \mathbf{R}^k \to \mathbf{R}$. We shall say that S is a *measurable set* if and only if χ_S is a measurable function. If χ_S is integrable we write

$$m(S) = \int \chi_S$$

and if χ_S is measurable, but not integrable, we write

$$m(S) = \infty;$$

$m(S)$, finite or infinite, is called the *(Lebesgue) measure* of S.

Many authors also say that S is *integrable* if and only if χ_S is integrable. We do not much care for this term because one might reasonably expect to be able to integrate an 'integrable set'. (Admittedly, one might also expect to be able to measure a 'measurable function'! We are no happier about this term but it is so firmly established in the literature that we cannot hope to alter it.) If χ_S is integrable we shall say that *S has finite measure* or that S is a *set of finite measure*.

There is no need to make the definition of Lebesgue measure depend on the idea of measurable function. We could equally well say that S *has finite measure*

$$m(S) = \int \chi_S$$

when χ_S is integrable and then say that S is *measurable* if $S \cap I$ has finite measure for every bounded interval I in $\mathbf{R}^k$. If S is measurable, according to this definition, and does not have finite measure, we naturally write

$$m(S) = \infty.$$

The equivalence of the two definitions is quite clear if we use the class (ii) of the previous section to truncate the characteristic function χ_S. For

$$\text{mid}\{-K\chi_I, \chi_S, K\chi_I\} = \chi_{S \cap I} \quad \text{if} \quad K \geq 1,$$
$$= K\chi_{S \cap I} \quad \text{if} \quad 0 \leq K \leq 1.$$

We should like to stress that the 'value' ∞ which we have introduced above is not to be interpreted as a number in the usual sense. As we said at the beginning of Chapter 1, we prefer not to augment the real line by adjoining points $-\infty, \infty$, and the definition we gave of unbounded intervals such as $[a, \infty), (-\infty, \infty)$ did not depend on the existence of such points. It could have been a dangerous luxury to allow ∞ as a possible value of our integrals (in particular when we regard them as linear operators). The danger of introducing ∞ in the context of measure is lessened by the fact that our measures are all positive and we are mostly involved in *adding* measures; nevertheless we must strictly avoid the 'difference' $\infty - \infty$, for example, in calculating the measure of a difference set $S \setminus T$. Incidentally, it is clear that the whole space $\mathbf{R}^k$ is measurable and contains bounded intervals of arbitrarily large measure, so that

$$m(\mathbf{R}^k) = \infty.$$

There is a natural extension of the notation introduced on p. 79. Let A be an arbitrary subset of $\mathbf{R}^k$. If $f: \mathbf{R}^k \to \mathbf{R}$ and $f\chi_A$ is integrable, then we write

$$\int f\chi_A = \int_A f$$

and say that f is *integrable on* A. The set of all such functions f is denoted by $L^1(A)$. In the same way we denote by $L^{\text{inc}}(A)$ the set of all functions f for which $f\chi_A \in L^{\text{inc}}(\mathbf{R}^k)$.

The following result lists the most fundamental properties of Lebesgue measure on $\mathbf{R}^k$.

Theorem 1. (i) *Let S, T be measurable sets; then so are $S \cup T$, $S \cap T$, $S \setminus T$, $S \triangle T$, and*
(a) $m(S) = 0$ *if and only if S is null;*
(b) $m(S) \geq m(T)$ *if $S \supset T$;*
(c) $m(S \cup T) + m(S \cap T) = m(S) + m(T)$.

(ii) *If $\{S_n\}$ is an increasing sequence of measurable sets and*

$$S = \bigcup_{n=1}^{\infty} S_n$$

then S is measurable and

$$m(S) = \lim m(S_n).$$

(iii) *If $\{S_n\}$ is a sequence of measurable sets and*

$$S = \bigcup_{n=1}^{\infty} S_n$$

then S is measurable and

$$m(S) \leq \sum_{n=1}^{\infty} m(S_n);$$

moreover, if the sets S_n are disjoint, then

$$m(S) = \sum_{n=1}^{\infty} m(S_n). \tag{1}$$

If we agree to write $a < \infty$ for any real number a then these results have a natural interpretation for sets of infinite measure. For example, in (i) (b) if $S \supset T$ and $m(T) = \infty$ then $m(S) = \infty$.

Proof. (i) The first part follows from Theorem 6.1.1 (ii) in view of the relations

$$\chi_{S \cup T} = \max\{\chi_S, \chi_T\},$$
$$\chi_{S \cap T} = \min\{\chi_S, \chi_T\},$$
$$\chi_{S \setminus T} = (\chi_S - \chi_T)^+,$$
$$\chi_{S \triangle T} = |\chi_S - \chi_T|.$$

(a) If S is null then $\chi_S = 0$ almost everywhere and so $\int \chi_S = 0$. If $\int \chi_S = 0$ then $\chi_S = 0$ almost everywhere by the Corollary to the Monotone Convergence Theorem. In other words, S is null.

(b) If we assume that S has finite measure, i.e. χ_S is integrable, then the inequality $\chi_S \geqslant \chi_T$ shows that χ_T is integrable and $\int \chi_S \geqslant \int \chi_T$. By the same token, if T has infinite measure then S must also have infinite measure.

(c) We have
$$\chi_{S \cup T} + \chi_{S \cap T} = \chi_S + \chi_T.$$

The various interpretations of this equation in case of infinite measures are quite obvious using (b).

(ii) This is immediate if we apply the Monotone Convergence Theorem to the increasing sequence $\{\chi_{S_n}\}$ which converges to χ_S. (If any $m(S_n) = \infty$ or if $\{m(S_n)\}$ diverges, then, of course, we have $m(S) = \infty$.)

(iii) Apply (ii) to the increasing sequence $\{T_n\}$, where
$$T_n = S_1 \cup \ldots \cup S_n.$$

According to Theorem 1 (iii), the union of a sequence of measurable sets is measurable. Bearing this in mind, the property (1) is often expressed by saying that Lebesgue measure is *countably additive* (or *σ-additive*). We note in passing that we can hardly expect Lebesgue measure to be more than countably additive: an interval $[a, b]$ in **R** with $a < b$ has non-zero measure but is the union of its individual points each of which has measure zero. Here, if you like, is a belated proof of the fact that the real numbers are not countable.

It was Lebesgue's considerable achievement in constructing a countably additive measure that opened the way to his powerful theory of integration. Earlier concepts of measure such as Jordan's 'content' based on ancient ideas of exhaustion going back as far as Archimedes, were only *finitely additive* in the sense that
$$m(S \cup T) = m(S) + m(T)$$
for any two disjoint measurable sets S, T, and these led to more limited theories of integration (cf. [22], Ch. 2).

We shall find it instructive to follow Lebesgue in defining measurable functions in terms of measure. Consider a function $f: \mathbf{R}^k \to \mathbf{R}$ and assume to begin with that f is *bounded*; thus there exist real numbers p, q for which
$$p \leqslant f(x) < q \quad \text{for all} \quad x \quad \text{(see Fig. 41).}$$

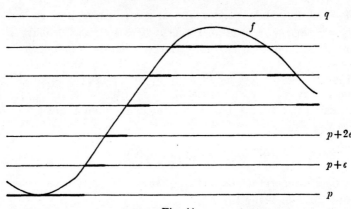

Fig. 41

If we now subdivide the interval $[p,q)$ into r disjoint intervals

$$I_i = [c_i, c_i + \epsilon) \quad (i = 1, \ldots, r)$$

each of length ϵ ($\epsilon = (q-p)/r$) we can guarantee a good approximation to f by means of the function

$$\phi = c_1 \chi_{S_1} + \ldots + c_r \chi_{S_r},$$

where $\quad S_i = \{x \in \mathbf{R}^k : c_i \leqslant f(x) < c_i + \epsilon\}.$

By this construction $\quad 0 \leqslant f(x) - \phi(x) < \epsilon$

for all x in $\mathbf{R}^k$. In other words we have found a function ϕ which takes only a finite number of values and which is *uniformly* near to our function f. We shall say that a function

$$\phi = c_1 \chi_{S_1} + \ldots + c_r \chi_{S_r}$$

is a *simple function* (or *generalised step function*) if S_i is measurable for $i = 1, 2, \ldots, r$.

For what functions f will this construction yield a simple function ϕ? According to Lebesgue a function $f: \mathbf{R}^k \to \mathbf{R}$ is *measurable* if and only if the sets

$$A_c = \{x \in \mathbf{R}^k : f(x) \geqslant c\}$$

are measurable for all c in $\mathbf{R}$. If this condition holds then, of course, every

$$A_c \backslash A_{c+\epsilon} = \{x \in \mathbf{R}^k : c \leqslant f(x) < c + \epsilon\}$$

is measurable and so we may apply the above construction to obtain a simple function ϕ which approximates the (bounded) function f uniformly on $\mathbf{R}^k$.

The equivalence of Lebesgue's definition and ours is contained in the following proposition.

Proposition 1. *The function $f: \mathbf{R}^k \to \mathbf{R}$ is measurable if and only if the sets*
$$A_c = \{x \in \mathbf{R}^k : f(x) \geq c\}$$
are measurable for all c in $\mathbf{R}$.

Proof. Assume that f is measurable. Then so are the functions f_n defined by
$$f_n(x) = n(\min\{f(x), c\} - \min\{f(x), c - (1/n)\}).$$
If $f(x) \geq c$, $f_n(x) = 1$, and if $f(x) < c$, $f_n(x) = 0$ for $n \geq N$ (depending on x). Thus
$$f_n \to \chi_{A_c}$$
which is therefore measurable by Theorem 6.1.1 (iii).

On the other hand, if A_c is measurable for every c in $\mathbf{R}$ then we may apply the above method of approximation to the bounded function
$$f_n = \operatorname{mid}\{-n, f, n - (1/n)\}$$
to obtain a simple function ϕ_n which satisfies
$$0 \leq f_n(x) - \phi_n(x) < 1/n \quad \text{for all } x.$$
(In this case we can take $p = -n$, $q = n$, $\epsilon = 1/n$, $r = 2n^2$.) Now the sequence $\{\phi_n\}$ of simple functions, each of which is obviously measurable, converges to f and so f is measurable. This completes the proof.

If we write
$$B_c = \{x \in \mathbf{R}^k : f(x) > c\}$$
then it is clear that
$$B_c = \bigcup_{n=1}^{\infty} A_{c+(1/n)}$$
so the measurability of the sets B_c follows from the measurability of the sets A_c by Theorem 1 (ii). In particular each set
$$A_c \setminus B_c = \{x \in \mathbf{R}^k : f(x) = c\}$$
is also measurable. From this it is now clear that the simple functions are precisely the measurable functions which assume only a finite number of values. Each simple function ϕ has a unique expression
$$\phi = c_1 \chi_{S_1} + \ldots + c_r \chi_{S_r}$$
where $c_1 < c_2 < \ldots < c_r$ are the *distinct* values assumed by ϕ, and
$$S_i = \{x \in \mathbf{R}^k : \phi(x) = c_i\}.$$
Note that zero may be one of these values, but in any case, the measurable sets S_i are disjoint and cover the whole of $\mathbf{R}^k$.

The above process of approximating f by means of 'horizontal strips' is often regarded as the hallmark of Lebesgue integration, and as such is contrasted with the more familiar idea of approximating f by means of 'vertical strips', i.e. by means of step functions (compare the Figs. on pp. 45, 128). The method of 'horizontal approximation' is only available after the notions of measure and measurable function have been defined; but granted that we have now constructed Lebesgue measure, the idea of 'horizontal approximation' allows us to express any measurable function as the limit *everywhere* of a sequence of simple functions, rather than merely as the limit *almost everywhere* of a sequence of step functions. This observation simplifies many arguments in the sequel.

Proposition 2. *A measurable function* $f: \mathbf{R}^k \to \mathbf{R}$ *may be expressed as the limit, everywhere, of a sequence of simple functions; if f is positive we may choose an increasing sequence of positive simple functions converging everywhere to f.*

Proof. The simple functions constructed in the proof of Proposition 1 satisfy all the conditions for the first part. When $f \geq 0$ we may adapt the construction slightly: for each integer $n \geq 1$, take $p = 0$, $q = n$, $\epsilon = 2^{-n}$, $r = n2^n$ and let $f_n = \min\{f, n - 2^{-n}\}$. The standard construction now gives a positive simple function ϕ_n which satisfies

$$0 \leq f_n - \phi_n < 2^{-n}$$

for all n. We have chosen $\epsilon = 2^{-n}$ (rather than $\epsilon = 1/n$ as above) so that $\phi_n \leq \phi_{n+1}$ for all n.

If two simple functions ϕ, ψ have expressions

$$\phi = c_1 \chi_{S_1} + \ldots + c_r \chi_{S_r},$$

$$\psi = d_1 \chi_{T_1} + \ldots + d_s \chi_{T_s}$$

in terms of their distinct values, then the intersections

$$S_i \cap T_j \quad (i = 1, \ldots, r; j = 1, \ldots, s)$$

are disjoint measurable sets which cover $\mathbf{R}^k$. It is clear that $\phi\psi$ is a simple function, and ϕ/ψ is a simple function provided $\psi(x) \neq 0$ for all x. Quite generally, if h is *any* real valued function whose domain of definition contains the ordered pairs (c_i, d_j) then the function θ defined by
$$\theta(x) = h(\phi(x), \psi(x)) \quad (x \in \mathbf{R}^k)$$

is again a simple function. To illustrate this idea we can prove the following useful result.

Proposition 3. *If $h: \mathbf{R}^2 \to \mathbf{R}$ is continuous and if $f, g: \mathbf{R}^k \to \mathbf{R}$ are measurable, then the composite function $F: \mathbf{R}^k \to \mathbf{R}$ defined by the equation*
$$F(x) = h(f(x), g(x))$$
for x in $\mathbf{R}^k$ is again measurable.

Proof. By Proposition 2, f, g are limits (everywhere) of sequences $\{\phi_n\}, \{\psi_n\}$ of simple functions. The function θ_n defined by
$$\theta_n(x) = h(\phi_n(x), \psi_n(x))$$
is also simple. By the continuity of h, $\theta_n(x) \to h(f(x), g(x))$ for all x, and so F is measurable.

Corollary. *If $f, g: \mathbf{R}^k \to \mathbf{R}$ are measurable, then their product fg is also measurable.*

Proof. Let $h(x, y) = xy$. (For an alternative proof see Ex. 6.1.11.)

Before we leave this section we must settle one outstanding question. Is it possible that *all* subsets of $\mathbf{R}^k$ are measurable and hence that *all* functions $f: \mathbf{R}^k \to \mathbf{R}$ are measurable? The answer to this is negative as we shall now see.

Define a relation $\sim$ on the real line $\mathbf{R}$ by setting
$$x \sim y$$
if and only if $x - y$ is rational. It is clear that $x \sim x$, that $x \sim y$ implies that $y \sim x$ and that $x \sim y$, $y \sim z$ imply $x \sim z$; in other words $\sim$ is an equivalence relation on $\mathbf{R}$. Let E be a set of points in $(0, 1)$ containing just one representative from each equivalence class. According to the *Axiom of Choice* such a choice of representatives is possible.† We can now show that E *is not Lebesgue measurable*.

If $x \in (0, 1)$ then $x \sim y$ for some y in E and so $x - y = r$ is a rational number in $(-1, 1)$. Thus
$$(0, 1) \subset \bigcup_{r \in (-1, 1)} (E + r) \subset (-1, 2), \tag{1}$$
where $E + r$ stands for the translated set $\{y + r : y \in E\}$. If r, s are rational numbers and $x \in (E + r) \cap (E + s)$, then $x = y + r = z + s$ where $y, z \in E$; thus $y - z = s - r$ and $y \sim z$. By the definition of E, this means that $y = z$ and so $r = s$. The translated sets $E + r$ for different rationals r in $(-1, 1)$ are therefore disjoint. Now suppose that E is Lebesgue

---

† In a comparatively recent paper [25] Solovay has proved that the Axiom of Choice (or equivalent) is indispensable for the construction of a non-Lebesgue-measurable set.

measurable. Then $m(E) = m(E+r)$ for any r and we may apply the σ-additivity of m to derive a contradiction: if $m(E) = 0$, the first inclusion of (1) would give $m((0, 1)) = 0$ which is impossible; and if $m(E) > 0$, the second inclusion of (1) would give $m((-1, 2)) = \infty$ which is likewise impossible. The conclusion is therefore that E is not Lebesgue measurable.

Exercises

1. If A, B are measurable subsets of $\mathbf{R}^k$ and $A \subset B$, show that
$$L^1(A) \supset L^1(B).$$

2. Prove a result similar to Theorem 1 (ii) for a decreasing sequence of measurable sets. Check your result for the case $S_n = (n, \infty)$ on $\mathbf{R}$.

3. Give an example of a non-measurable function $f: \mathbf{R} \to \mathbf{R}$ and use f to construct a non-measurable function $g: \mathbf{R}^2 \to \mathbf{R}$.

4. (i) Give an example of a non-measurable function f whose square is measurable.
(ii) If f^2 is measurable and $\{x : f(x) > 0\}$ is measurable, show that f is measurable. (Cf. Ex. 6.1.4.)

5. Let $f: \mathbf{R}^k \to \mathbf{R}$ and define $\dfrac{1}{f}: \mathbf{R}^k \to \mathbf{R}$ by
$$\frac{1}{f}(x) = \frac{1}{f(x)} \quad \text{if} \quad f(x) \neq 0,$$
$$= 0 \quad \text{if} \quad f(x) = 0.$$
If f is measurable show that $1/f$ is measurable by using
(i) Lebesgue's definition in terms of the sets A_c (p. 128),
(ii) a sequence of simple functions converging to f.

6. Let $f: \mathbf{R}^k \to \mathbf{R}$.
(i) Show that f is measurable if and only if $f^{-1}(I)$ is measurable for every bounded interval I in $\mathbf{R}$. (The *inverse image* $f^{-1}(I)$ consists of all points x in $\mathbf{R}^k$ for which $f(x) \in I$.)
(ii) Show that f is measurable if and only if $\phi \circ f$ is simple for every step function ϕ on $\mathbf{R}$. (The *composite function* $\phi \circ f$ is defined by
$$(\phi \circ f)(x) = \phi(f(x))$$
for all x in $\mathbf{R}^k$.)

7. (i) Let $\phi = c_1 \chi_{S_1} + \ldots + c_r \chi_{S_r}$ be the unique expression for a simple function ϕ in terms of its distinct *non-zero* values $c_1, \ldots, c_r$. Show that ϕ is integrable if and only if $S_1, \ldots, S_r$ all have finite measure, and then
$$\int \phi = c_1 m(S_1) + \ldots + c_r m(S_r).$$

(ii) Let $f: \mathbf{R}^k \to \mathbf{R}$ be a positive measurable function and let $\{\phi_n\}$ be the increasing sequence of positive simple functions constructed in Proposition 2. Show that $f \in L^1(\mathbf{R}^k)$ if and only if the sequence $\{\int \phi_n\}$ is convergent, and then
$$\int \phi_n \to \int f.$$
(This gives an alternative definition of $\int f$ for positive f. For an arbitrary measurable function f it may be applied separately to f^+, f^- and $\int f$ defined as $\int f^+ - \int f^-$.)

8. Let f be a measurable function on $\mathbf{R}$.
(i) If
$$g(x) = 1/2^n \quad \text{for} \quad n-1 \leq |x| < n$$
($n = 1, 2, \ldots$), draw the graph of g and show that $g \in L^1(\mathbf{R})$.
(ii) Use Proposition 3 to show that
$$h = \frac{f^+ g}{f^+ + g}$$
is integrable.
(iii) Let $\{\phi_n\}$ be a sequence of step functions converging almost everywhere to h and define
$$\psi_n = \frac{g \phi_n}{g - \phi_n}$$
(equal to zero when $g = \phi_n$). Show that $\{\psi_n\}$ is a sequence of step functions converging almost everywhere to f^+.
(iv) Hence show that f is the limit almost everywhere of a sequence of step functions.

9. Generalise Ex. 8 to the case of a measurable function f on $\mathbf{R}^k$.

10. Let $A \subset \mathbf{R}^l$, $B \subset \mathbf{R}^m$ and define
$$A \times B = \{(a,b) : a \in A, b \in B\}.$$
We may think of $A \times B$ as a subset of $\mathbf{R}^k$ ($k = l+m$).
(i) If N is null in $\mathbf{R}^l$ show that $N \times \mathbf{R}^m$ is null in $\mathbf{R}^k$.
(ii) If A, B are measurable subsets of $\mathbf{R}^l$, $\mathbf{R}^m$, respectively, show that $A \times B$ is a measurable subset of $\mathbf{R}^k$. Under what circumstances can we say that
$$m(A \times B) = m(A) m(B)?$$

11. Let f be a positive measurable function on $\mathbf{R}^k$ and define the *ordinate sets* in $\mathbf{R}^{k+1}$:
$$P_f = \{(x,y) : 0 \leq y < f(x)\},$$
$$Q_f = \{(x,y) : 0 \leq y \leq f(x)\}.$$
Note that P_f is 'open at the top' and Q_f is 'closed at the top'.

(i) Show that the ordinate set P_χ is $A \times [0,1]$ and find Q_{χ_A}. (This is one good reason for preferring P_f to Q_f.)

(ii) Deduce that the ordinate set P_ϕ of a positive simple function ϕ is measurable (see Ex. 10).

(iii) Use Proposition 2 to show that P_f is measurable.

(iv) By considering the ordinate sets $P_{f+(1/n)}$ ($n = 1, 2, \ldots$), show that Q_f is measurable.

(v) Show that f is integrable if and only if the ordinate set P_f has finite measure, and that $\int f = m(P_f)$ in this case. (This makes precise, and extends to $\mathbf{R}^k$, the famous definition of $\int f$ as the 'area under the graph'. It applies to positive f but can be extended to an arbitrary function f in L^1 by considering f^+, f^- separately.)

12. (i) Let f be a positive increasing function on the closed interval $[a, b]$. For given $\epsilon > 0$, adapt the construction on p. 128 to find a positive increasing step function ϕ on $[a, b]$ which satisfies

$$0 \leqslant f(x) - \phi(x) < \epsilon$$

for all x in $[a, b]$.

(ii) Use (i) to prove **Bonnet's Mean Value Theorem**. *If f is positive and increasing on $[a, b]$ and $g \in L^1[a, b]$, then $fg \in L^1[a, b]$ and there is a point c in $[a, b]$ for which*

$$\int_a^b fg = f(b) \int_c^b g.$$

N.B. The left hand side of this equation is independent of the value of f at the particular point b. In practice it is often convenient to assume that $f(b) = f(b-0)$ (remember that f is increasing). Of course, we shall expect the point c to depend on the value chosen for $f(b)$.

6.3 The geometry of Lebesgue measure

According to one very famous definition, a *geometry* on a space S consists of the properties of S that are invariant under a group of transformations of S. This definition was given, in a much more precise form, by Felix Klein in his inaugural lecture when he accepted the Chair of Mathematics at Erlangen University in 1872, at the tender age of 23; it has been known as the '*Erlanger Programm*' ever since. In this section we shall be interested in three kinds of geometry on the space $\mathbf{R}^k$: *Euclidean geometry*, *affine geometry*, and the standard *topology* defined in terms of the Euclidean distance; and we shall look at the effect of their transformations on the measure we have constructed.

It is convenient to begin with some general remarks about mappings. (The terms 'mapping', 'transformation', 'function' are synonymous, but it is common practice to use 'transformation' or 'mapping' in a geometrical context.) Let X, Y be two non-empty sets and T a mapping of X into Y. If $A \subset X$ then

$$TA = \{Ta : a \in A\}.$$

If $B \subset Y$ then $\quad T^{-1}B = \{x \in X : Tx \in B\};$

it is not assumed that $B \subset TX$. We shall say that T is *onto* Y if $TX = Y$ and that T is *one-one* if $T^{-1}\{y\}$ consists of exactly one element for each y in TX. The reader will easily verify that T is a one–one mapping of X onto Y if and only if there exists a mapping T^{-1} of Y into X satisfying

$$T^{-1}(Tx) = x, \quad T(T^{-1}y) = y$$

for all x in X, y in Y. In this case T^{-1} is called the *inverse* of T.

The fundamental sets in any topological space are the *open sets*. For $\mathbf{R}^k$ we define these in terms of our given distance function. A set A of points contained in $\mathbf{R}^k$ will be called *open* if to any point a of A there corresponds a ball $|x-a| < \delta$ ($\delta > 0$) which lies entirely in A. We note in particular that the whole space $\mathbf{R}^k$ is open, and also the empty set (by default, as it were, because there are no points for which the above criterion can fail). A set B is *closed* if and only if the complementary set $\mathbf{R}^k \backslash B$ is open. It is clear that the 'open ball' $|x-a| < \delta$ is indeed open and the 'closed ball' $|x-a| \leq \delta$ is indeed closed. There exist plenty of subsets of $\mathbf{R}^k$ which are neither open nor closed – for example, the half-open intervals of type $[\,,\,)^k$.

The reader who is unfamiliar with these elementary notions of topology may find it helpful to read the Appendix. In particular, the bounded closed subsets of $\mathbf{R}^k$ are there identified with the *compact* subsets of $\mathbf{R}^k$ (Appendix, Theorem 7). In the present chapter, when we are discussing subsets of $\mathbf{R}^k$, we may regard 'compact' as a shorthand for 'bounded and closed'.

The following simple result has far reaching consequences, as many of the sets we encounter in $\mathbf{R}^k$ are obtained by taking (countable) unions and intersections of open or closed sets.

Proposition 1. *All open sets in $\mathbf{R}^k$ and all closed sets in $\mathbf{R}^k$ are measurable. A compact subset of $\mathbf{R}^k$ has finite measure.*

Proof. Let A be an open set in $\mathbf{R}^k$. Consider the collection $\mathscr{S}_0$ of unit cubes defined by the inequalities

$$m_i \leq x_i < m_i + 1,$$

where each m_i $(i = 1, 2, ..., k)$ is an integer; these cubes are disjoint and cover the whole of $\mathbf{R}^k$. As there are only a finite number of them in each cube I_N defined by $|x_i| \leqslant N$ $(i = 1, 2, ..., k)$, we may apply Proposition 2.1.1 to show that $\mathscr{S}_0$ is countable. Now let S_0 be the union of the unit cubes of $\mathscr{S}_0$ that are contained in A.

By successive halving of the sides we may subdivide each of the cubes of $\mathscr{S}_0$ into 2^{kn} equal cubes: in this way we obtain a countable collection $\mathscr{S}_n$ of cubes of side 2^{-n}. Let S_n be the union of the cubes of $\mathscr{S}_n$ that are contained in A (Fig. 42).

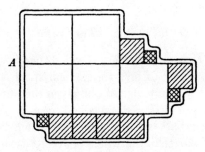

Fig. 42. (Blank squares belong to S_1; lightly shaded squares belong to $S_2 \backslash S_1$; doubly shaded squares belong to $S_3 \backslash S_2$.)

Each of the sets S_n is the union of a countable collection of cubes and as such is certainly measurable. By our construction the sequence $\{S_n\}$ is increasing, so the measurability of A will follow by Theorem 6.2.1 (ii) if we can show that $A = \bigcup S_n$. Let a be an arbitrary point of A. As A is open in $\mathbf{R}^k$, there is a radius $\delta > 0$ such that the open ball defined by $|x - a| < \delta$ is contained in A. If n is large enough to ensure that the diagonal of each cube in $\mathscr{S}_n$ is less than δ then there is one (and only one) cube of S_n containing the point a (see Fig. 43). Each S_n is contained in A by construction and so

$$A = \bigcup S_n.$$

The measurability of a closed set now follows at once by Theorem 6.2.1 (i). A compact subset of $\mathbf{R}^k$ may always be enclosed in a bounded interval and so a compact subset has finite measure.

There is a modification of the above proof that is important in the sequel. We may define X_n as the union of the cubes of $\mathscr{S}_n$ whose *closures* are contained in A. In general X_n is smaller than S_n, but exactly the same argument shows that $A = \bigcup X_n$. Each difference

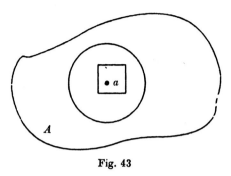

Fig. 43

$X_n \backslash X_{n-1}$ is the union of cubes in $\mathscr{S}_n$. This gives the following useful result.

Proposition 2. *An open set A in $\mathbf{R}^k$ may be expressed as the union of a sequence of disjoint cubes whose closures are contained in A.*

The transformations studied in Euclidean geometry are the ones which preserve length: an *isometry* of $\mathbf{R}^k$ is a mapping T of $\mathbf{R}^k$ into $\mathbf{R}^k$ satisfying
$$|Tx - Ty| = |x - y|$$
for all x, y in $\mathbf{R}^k$. It is clear that such an isometry T is *one-one* and we shall see presently that T is *onto* $\mathbf{R}^k$.

We need one or two elementary facts about the space $\mathbf{R}^k$. The *scalar product* $x.y$ is defined by
$$x.y = x_1 y_1 + \ldots + x_k y_k.$$
Thus $x.x = |x|^2$.

Proposition 3 (*Schwarz' Inequality*).
$$|x.y| \leqslant |x||y| \quad (x, y \in \mathbf{R}^k). \tag{1}$$

Proof. The case $x = 0$ is trivial, so we assume that $x \neq 0$. Now
$$|tx - y|^2 = (tx - y).(tx - y) = t^2 x.x - 2tx.y + y.y$$
is positive for all real numbers t. In particular $t = x.y/x.x$ yields
$$-(x.y)^2/x.x + y.y \geqslant 0$$
from which (1) follows.

This argument also shows that we can only have equality in (1) if $y = tx$ for some real t (on the assumption that $x \neq 0$).

Proposition 4 (*The Triangle Inequality*).

$$|x+y| \leq |x|+|y| \quad (x, y \in \mathbf{R}^k). \tag{2}$$

Proof.
$$\begin{aligned}|x+y|^2 &= (x+y).(x+y) \\ &= x.x + 2x.y + y.y \\ &\leq |x|^2 + 2|x||y| + |y|^2\end{aligned}$$

by Schwarz' Inequality. If we also assume that $x \neq 0$, equality in (2) implies that $y = tx$ for some $t \geq 0$.

Corollary.

$$|a-b| \leq |a-c| + |c-b| \quad (a, b, c \in \mathbf{R}^k).$$

If a, b, c are distinct then the equation

$$|a-b| = |a-c| + |c-b|$$

implies that $c - b = t(a-c)$, where $t > 0$. In other words

$$c = \frac{ta+b}{t+1},$$

where $t > 0$. The *(straight line) segment* joining a to b may be defined as the set of all points $(1-s)a + sb$ for $0 \leq s \leq 1$. It is clear that the above point c is on the segment joining a to b and satisfies

$$|c-b| = t|a-c|$$

(see Fig. 44). These remarks give substance to the familiar statement, 'A straight line is the shortest distance between two points.' In our context they have the more important function of showing that the collinearity of three points is expressible in terms of distance only, and so *collinearity is preserved by isometries*. We can make a rather stronger statement as follows.

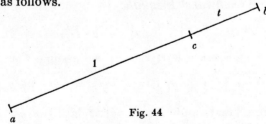

Fig. 44

Theorem 1. *An isometry of $\mathbf{R}^k$ which does not move the origin is a linear mapping.*

Proof. Let T be an isometry of $\mathbf{R}^k$ satisfying $T0 = 0$. The above argument applied to the points $0, a, ta$ shows that they map to

$0, Ta, tTa$. (The cases $t < 0, 0 < t < 1, 1 < t$ may be treated separately. See Ex. 1.) Thus
$$T(ta) = tTa \quad (t \in \mathbf{R}). \tag{3}$$

Moreover the mid point $\tfrac{1}{2}(a+b)$ maps to the mid point $\tfrac{1}{2}(Ta+Tb)$ and so, using (3),
$$T(a+b) = Ta+Tb. \tag{4}$$

These two equations show that T is a linear mapping.

An *affine mapping* T of $\mathbf{R}^k$ is defined by
$$Tx = a + Lx \quad (x \in \mathbf{R}^k),$$
where $a \in \mathbf{R}^k$ and $L: \mathbf{R}^k \to \mathbf{R}^k$ is a linear mapping.

Corollary. *An isometry of $\mathbf{R}^k$ is an affine mapping.*

Proof. Let T be an isometry of $\mathbf{R}^k$; set $a = T0$ and $Lx = Tx - a$. It is clear that L is an isometry and $L0 = 0$.

Our task is now to find the effect on Lebesgue measure of an arbitrary affine mapping; in view of the invariance of Lebesgue measure under translations (Ex. 4.2.5) it is enough to consider linear mappings.

It is a familiar fact that a linear mapping $L: \mathbf{R}^k \to \mathbf{R}^k$ may be described by a $(k \times k)$ real matrix: if $y = Lx$ then
$$\begin{aligned} y_1 &= c_{11}x_1 + c_{12}x_2 + \ldots + c_{1k}x_k, \\ y_2 &= c_{21}x_1 + c_{22}x_2 + \ldots + c_{2k}x_k, \\ &\ldots \\ y_k &= c_{k1}x_1 + c_{k2}x_2 + \ldots + c_{kk}x_k. \end{aligned} \tag{5}$$

The $(k \times k)$ matrix
$$\begin{pmatrix} c_{11} & c_{12} & \ldots & c_{1k} \\ c_{21} & c_{22} & \ldots & c_{2k} \\ \ldots \\ c_{k1} & c_{k2} & \ldots & c_{kk} \end{pmatrix}$$
is often written (c_{ij}), or C, for short.

There is a one-one correspondence between $(k \times k)$ real matrices C and linear mappings $L: \mathbf{R}^k \to \mathbf{R}^k$ defined by the equations (5). This correspondence is further enriched by defining addition and multiplication of matrices to correspond exactly with the addition and multiplication of linear mappings. Thus
$$(a_{ij}) + (b_{ij}) = (a_{ij} + b_{ij}),$$
$$(a_{ij})(b_{ij}) = (c_{ij}),$$

where
$$c_{ij} = \sum_{s=1}^{k} a_{is} b_{sj}.$$

The *unit matrix*
$$I = \begin{pmatrix} 1 & 0 & 0 & \ldots & 0 \\ 0 & 1 & 0 & \ldots & 0 \\ \ldots & & & & \\ 0 & 0 & 0 & \ldots & 1 \end{pmatrix}$$

describes the identity mapping which maps x on x for all x in $\mathbf{R}^k$.

The matrix P is *invertible* if there is a matrix Q satisfying
$$PQ = QP = I.$$

We shall be particularly interested in two kinds of matrix.

(S) For $i \neq j$ and t in $\mathbf{R}$ let $S_{ij}(t)$ denote the matrix obtained from I by adding t times the j-th row to the i-th row. The corresponding affine mapping is a '*shear in the (i,j) plane parallel to the i-axis*'.

(R) For $i \neq j$ let R_{ij} denote the matrix obtained from I by interchanging the i-th and j-th rows. The corresponding affine mapping is a '*reflection in the hyperplane $x_i = x_j$*'.

Matrices of types (S), (R) are often called *elementary matrices*. The paragraph which follows and the proof of Lemma 1 assume some familiarity with elementary row and column operations. The reader is expected to fill in a good many details for himself – these ideas are to be found in any basic text on linear algebra and matrices.

For any $(k \times k)$ matrix C, $S_{ij}(t) C$ is obtained from C by adding t times the j-th row to the i-th row, and $C S_{ij}(t)$ by adding t times the i-th column to the j-th column. Also $R_{ij} C$ is obtained from C by interchanging the i-th and j-th rows and $C R_{ij}$ by interchanging the i-th and j-th columns.

We can now prove an elementary algebraic lemma which greatly simplifies some of our subsequent proofs.

Lemma 1. *Let C be a $(k \times k)$ real matrix. Then there exist invertible matrices P, Q each of which is the product of elementary matrices (of types (S) or (R)) such that PCQ is a diagonal matrix.*

Proof. If $C = 0$, the zero matrix, there is nothing to prove. Assume that $C \neq 0$ and bring a non-zero element into the $(1, 1)$ position, if necessary, by suitable interchanges of rows and columns. By adding suitable multiples of the new first column to the others we obtain zeros in the $(1, 2), \ldots, (1, k)$ positions, and by similar operations on the rows, obtain zeros in the $(2, 1), \ldots, (k, 1)$ positions. We may now apply the same argument to the elements which are not in the first row or column and so produce a diagonal matrix after at most $(k-1)$ such reductions.

All these elementary row and column operations are invertible and are given by pre- or post-multiplication by elementary matrices; these matrices may be multiplied together to give the required matrices P, Q.

Let $L: \mathbf{R}^k \to \mathbf{R}^k$ be an arbitrary linear mapping whose matrix C satisfies
$$PCQ = \begin{pmatrix} d_1 & & & \\ & d_2 & & \\ & & \ddots & \\ & & & d_k \end{pmatrix}$$
as in Lemma 1. The first point to note is that C is invertible if and only if PCQ is invertible, which in turn is true if and only if $d_1 d_2 \ldots d_k \neq 0$. In fact L is *one-one* if and only if $d_1 d_2 \ldots d_k \neq 0$ and L is onto $\mathbf{R}^k$ if and only if $d_1 d_2 \ldots d_k \neq 0$; so that linear mappings $L: \mathbf{R}^k \to \mathbf{R}^k$ have the special property that they are invertible if and only if they are one-one, if and only if they are onto $\mathbf{R}^k$. This gives a belated proof that *an isometry of $\mathbf{R}^k$ must be onto $\mathbf{R}^k$*.

Suppose now that $T: \mathbf{R}^k \to \mathbf{R}^k$ is an affine mapping. We are interested in the measure of TI for any bounded interval I. It is probably simplest to note once and for all the following result.

Proposition 5. *Let $T: \mathbf{R}^k \to \mathbf{R}^k$ be a continuous mapping. Then TI has finite measure for any bounded interval I in $\mathbf{R}^k$.*

Continuity of T is defined exactly as in §4.2, or more fully in the Appendix.

Proof. This is clear in the first place if I is compact because TI is compact in this case (Appendix, Theorem 8) and we may apply Proposition 1. In general we may express a bounded interval I as the union of an increasing sequence $\{I_n\}$ of compact intervals. (For example, in $\mathbf{R}^1$, the interval $(0, 1]$ is the union of the compact intervals $[1/n, 1]$ $(n = 1, 2, \ldots)$; this idea clearly generalises to $\mathbf{R}^k$.) But this gives
$$TI = \bigcup TI_n$$
(Ex. 5) and so TI has finite measure (not greater than $m(T\bar{I})$ where $\bar{I}$ is the closure of I).

If the linear mapping L has matrix (c_{ij}) then we may apply Schwartz' Inequality to each y_i to see that
$$y_i^2 \leqslant (c_{i1}^2 + \ldots + c_{ik}^2) |x|^2$$

and hence
$$|y| \leq M|x|, \tag{6}$$
where
$$M = (\sum_{i,j} c_{ij}^2)^{\frac{1}{2}}. \tag{7}$$

This inequality shows at once that the linear mapping L is continuous (cf. the Lipschitz condition of Ex. 3.5.6). Thus the affine mapping T given by $Tx = a + Lx$ is also continuous and TI has finite measure for any bounded interval I.

Theorem 2. *Let $T: \mathbf{R}^k \to \mathbf{R}^k$ be an affine mapping and let E be an arbitrary subset of $\mathbf{R}^k$ of finite measure; then TE has finite measure*
$$m(TE) = \Delta_T m(E), \tag{8}$$
where Δ_T is a constant (depending only on T). More generally,
$$\int g = \Delta_T \int f \tag{9}$$
for any f in L^1, where f, g are related by the equation
$$g(Tx) = f(x) \tag{10}$$
for all x in $\mathbf{R}^k$ and g vanishes outside $T\mathbf{R}^k$.

Proof. First of all, if L is a linear mapping with elementary matrix of type (S) or (R), then we verify at once, using Fubini's Theorem, that $m(LI) = m(I)$ for any bounded interval I. This is illustrated by the 'shear' in $\mathbf{R}^2$:
$$y_1 = x_1 + tx_2,$$
$$y_2 = x_2 \quad \text{(see Fig. 45)}.$$

Fubini's Theorem allows us to calculate the area (once we are sure that it exists) by integrating the cross-sections parallel to the 1-axis; from this it is clear that $m(LI) = m(I)$. The general shear reduces to this one immediately.

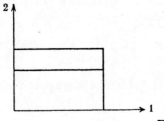
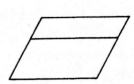

Fig. 45

6.3] GEOMETRY OF MEASURE

Further, if L has diagonal matrix, with diagonal entries $d_1, \ldots, d_k$, then it is equally clear that

$$m(LI) = |d_1 d_2 \ldots d_k| m(I).$$

Suppose now that T is a particular affine mapping for which there is a constant Δ_T satisfying

$$m(TI) = \Delta_T m(I) \tag{11}$$

for all bounded intervals I. (For example, T could be one of the special mappings just mentioned above.) The main part of the proof consists in showing that the theorem holds for such a mapping T.

Equation (11) may be regarded as a special case of (9) with $f = \chi_I$. This extends by linearity to the case where f is a step function on $\mathbf{R}^k$. Now if $f \in L^{\text{inc}}$ there is an increasing sequence $\{\phi_n\}$ of step functions converging to f outside a null set S, say. The 'distorted step functions' ψ_n defined by

$$\psi_n(Tx) = \phi_n(x)$$

(which are constant on 'distorted intervals' TI) provide an increasing sequence which converges to g outside the set TS. Assuming for the moment that TS is null, it follows by the Monotone Convergence Theorem that (9) holds for any f in L^{inc}. The final extension to L^1 is immediate on taking differences, and gives (8) when we substitute $f = \chi_E$.

The proof that TS is null is reminiscent of ideas we have seen before in part (iii) of the proof of Fubini's Theorem (§4.3) and also in the proof of the formula for Integration by Substitution (Proposition 5.1.3); we shall meet these ideas several times in the sequel. As S is null we can find an increasing sequence $\{\theta_n\}$ of step functions diverging on S but with $\int \theta_n \leq K$. The corresponding 'distorted step functions' σ_n defined by

$$\sigma_n(Tx) = \theta_n(x)$$

are the terms of an increasing sequence which diverges on TS and has

$$\int \sigma_n = \Delta_T \int \theta_n \leq \Delta_T K.$$

The Monotone Convergence Theorem now shows that $\{\sigma_n\}$ is convergent almost everywhere and so TS is null.

The rest of the proof is easy. We have verified (11) for the special linear mappings listed at the beginning of the proof and so (8) holds for each of them. But it is now clear that (8) is satisfied for finite products of linear mappings whose matrices are either elementary or diagonal. The theorem therefore follows from Lemma 1.

If T_1, T_2 are any two affine mappings on $\mathbf{R}^k$ then (8) shows that

$$\Delta_{T_1 T_2} = \Delta_{T_1} \Delta_{T_2}. \tag{12}$$

Moreover $\Delta_T = 1$ for a shear or a reflection and $\Delta_T = |d_1 d_2 \ldots d_k|$ for a linear mapping with diagonal matrix as above. It is now clear that *an affine mapping T is invertible if and only if $\Delta_T \neq 0$.*

Those who are familiar with *determinants* will recognise that

$$\Delta_T = |\det C|, \tag{13}$$

where $Tx = a + Lx$ and C is the matrix of L. In fact Lemma 1 corresponds to a standard method of evaluating a determinant by elementary row and column operations, the only point of difference being that an interchange of rows or columns produces a change of sign in the determinant. Our treatment makes no use of determinants, and in particular makes no use of the multiplication rule for determinants to establish (12).

In order to calculate the constant Δ_T for a given T we may not need to do any algebra. For example if T is an isometry we may assume without loss of generality that $T0 = 0$ and take for A the unit closed ball centre 0 defined by $|x| \leq 1$. Clearly $TA = A$ and $m(A) > 0$ so that $\Delta_T = 1$. This proves the following important result which shows, at last, that the measure which we have defined in $\mathbf{R}^k$ does not depend on the very special frame of reference.

Corollary. *Lebesgue measure is invariant under isometries.*

To close this section we must say a word about the topological transformations of $\mathbf{R}^k$ and their effect on Lebesgue measure. Let A, B be non-empty subsets of $\mathbf{R}^k$. A one-one mapping T of A onto B which is continuous and has a continuous inverse T^{-1} is called a *topological transformation*, or a *homeomorphism*, of A onto B. In view of Propositions 1 and 2 it may come as a surprise that homeomorphisms have no respect whatsoever for Lebesgue measure! Even on the real line, where there hardly seems to be enough room for much distortion, we can find a homeomorphism that maps a set of measure zero onto a set of measure 1 and maps a measurable set onto a non-measurable set. Some details are given in Exx. 6–8.

For a general homeomorphism $T: \mathbf{R}^k \to \mathbf{R}^k$ and a measurable set E the best we can do is to find two 'good' sets A, B such that

$$A \subset E \subset B, \quad m(B \setminus A) = 0$$

and TA, TB are measurable, even though we may have

$$m(TA) < m(TB).$$

These 'good' sets are the *Borel sets* discussed in the sequel.

On the other hand, it is encouraging to know that a homeomorphism T which possesses a *density function* Δ_T, analogous to the constant Δ_T of Theorem 2, has almost all the properties that we could hope for; this is the substance of the next section.

Exercises

1. Fill in the details of the first part of the proof of Theorem 1.

2. What transformation T of $\mathbf{R}^2$ is described by the matrix

$$P = \begin{pmatrix} \cos\theta & -\sin\theta \\ \sin\theta & \cos\theta \end{pmatrix}?$$

Calculate Δ_T. Find the inverse matrix of P.

3. If the linear mapping L has matrix

$$\begin{pmatrix} a & b \\ c & d \end{pmatrix}$$

show that $\Delta_L = |ad - bc|$.

4. The matrix

$$\begin{pmatrix} 1 & 2 & 1 \\ 0 & -1 & 1 \\ 1 & 1 & 3 \end{pmatrix}$$

defines a linear mapping L. Use Lemma 1 to calculate Δ_L.

5. Let T be a mapping of X into Y and $\{A_n\}$ a sequence of subsets of X. If

$$A = \bigcup A_n.$$

show that

$$TA = \bigcup TA_n.$$

Would the same be true if $\bigcup$ were replaced by $\bigcap$?

6. Let F be the Cantor function described in Ex. 3.5.10 and define

$$T(x) = x + F(x)$$

for x in $[0, 1]$. Show that T is continuous and strictly increasing, and hence that T is a homeomorphism of $[0, 1]$ onto $[0, 2]$. Show that T maps Cantor's ternary set (§ 2.3), of measure zero, onto a set of measure 1.

7. Let F be a subset of $(0, 1)$ with Lebesgue measure $m(F) > 0$. Adapt the last paragraph of § 6.2 as follows: let E be a set of representatives in F

of those equivalence classes that have non-empty intersection with F; show that
$$F \subset \bigcup_{r \in (-1,1)} (E+r) \subset (-1,2)$$
(r rational) and deduce that the subset E of F is not Lebesgue measurable.

8. Use Exx. 6, 7 to find a homeomorphism which maps a measurable set onto a non-measurable set.

6.4 Transformation of integrals

Throughout the present section U will be an *open* subset of $\mathbf{R}^k$ and T a *one-one* continuous mapping of U into $\mathbf{R}^k$. The proof of Proposition 6.3.5 shows that TI has finite measure for any bounded interval I whose closure $\bar{I}$ is contained in U. We have just seen at the end of the previous section how badly behaved T can be. But suppose that there is a positive function $h\colon U \to \mathbf{R}$ such that

$$m(TI) = \int_I h \tag{1}$$

for any bounded interval I with $\bar{I} \subset U$. In this case we shall say that h is a *density function* for T. We stress the fact that this definition only involves $m(TI)$ for bounded intervals I (with $\bar{I} \subset U$) and not $m(TE)$ for arbitrary measurable subsets E of U.

Theorem 6.3.2 shows that an affine mapping T has constant density function Δ_T. A less obvious example is provided by the *polar transformation*. Let U be the open strip of $\mathbf{R}^2$ consisting of points (r, θ) with $0 < r$, $0 < \theta < 2\pi$ and define $T(r, \theta) = (x, y)$ where

$$x = r\cos\theta, \quad y = r\sin\theta.$$

The image TU is the open set $\mathbf{R}^2 \setminus \{(x, 0) : x \geq 0\}$, i.e. the plane $\mathbf{R}^2$ with the positive real axis removed. We take it as a well known property of the trigonometric functions that this transformation is one-one; the continuity of T follows from the continuity of the cosine and sine functions. If I is the interval in U defined by the inequalities

$$r_1 \leq r \leq r_2, \quad \theta_1 \leq \theta \leq \theta_2$$

then TI has area $\quad \tfrac{1}{2}(r_2^2 - r_1^2)(\theta_2 - \theta_1)$
(see Fig. 46).

This can be verified, at a very elementary level, using only the formula πr^2 for the area of a circular disk. In terms of equation (1) it means that
$$h(r, \theta) = r$$

6.4] TRANSFORMATIONS 147

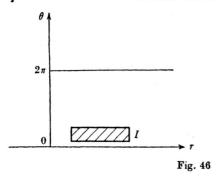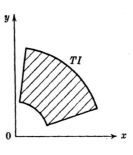

Fig. 46

defines a density function h for T because

$$m(TI) = \int_{\theta_1}^{\theta_2} \left(\int_{r_1}^{r_2} r\,dr \right) d\theta,$$

and, in this case, the boundary points of I or of TI contribute nothing to the measure (Ex. 4.2.8). The same kind of elementary verification yields density functions for the cylindrical polar and spherical polar transformations. See Exx. 1, 2, 3 for details.

The following fundamental theorem shows that the existence of a density function h guarantees the good behaviour of T.

Theorem 1. *Let U be an open subset of $\mathbf{R}^k$ and $T: U \to \mathbf{R}^k$ a one-one continuous mapping with density function h. Then TE is (Lebesgue) measurable for any (Lebesgue) measurable subset E of U,*

$$m(TE) = \int_E h, \tag{2}$$

and

$$\int_V g(y)\,dy = \int_U g(Tx)\,h(x)\,dx \tag{3}$$

where $V = TU$.

Equation (3) is to be interpreted in the most generous possible way: if one integral exists, then so does the other, and they are equal. Similarly, equation (2) means that TE has finite measure if and only if $h \in L^1(E)$ and the measure is then equal to $\int_E h$.

The proof that follows is rather long and involved. The easier first part, which establishes (3) on the assumption that $g \in L^1(V)$, is all that is usually required in practice. The strategy of this first part is briefly as follows: (2) is established for open subsets of U, and then (3) is established for the cases where g is successively the characteristic function of a bounded open set, the characteristic function of a bounded interval, a step function, an element of $L^{\text{inc}}(V)$, and finally an element of $L^1(V)$.

Proof. Let A be an open subset of U. Then A may be expressed as the union of a sequence $\{C_n\}$ of disjoint cubes with $\bar{C}_n \subset A$ (Proposition 6.3.2) and TA is the union of the *disjoint* sets TC_n. (Here we use the fact that T is one–one.) As h is a density function for T,

$$m(TC_n) = \int_{C_n} h$$

so that
$$m(TA) = \Sigma \int_{C_n} h$$

by the countable additivity of measure. This series is convergent if and only if $m(TA)$ is finite. The Monotone Convergence Theorem now gives

$$m(TA) = \int_A h,$$

where this equation is interpreted as meaning that $m(TA) = \infty$ if $h \notin L^1(A)$. As a particular case we see that $V = TU$ *is measurable.*

If B is an arbitrary bounded open subset of $\mathbf{R}^k$, then

$$T^{-1}B = A,$$

say, is an open subset of U (see Fig. 47). (Here we use the continuity of T and Proposition 3 of the Appendix.) Thus $TA = B \cap V$ has finite measure $\int_A h$. In other words (3) holds in the special case $g = \chi_B$.

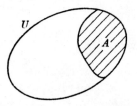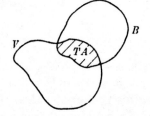

Fig. 47

Now any bounded interval J in $\mathbf{R}^k$ may be expressed as the difference of two bounded open subsets of $\mathbf{R}^k$ one of which contains the other, so that (3) holds in the special case $g = \chi_J$; the extension to step functions is immediate by linearity.

(The paragraph which follows should be compared with the proofs of Fubini's Theorem 4.3.1, Proposition 5.1.3 and Theorem 6.3.2.) Sup-

pose that $g \in L^{\text{inc}}(V)$ and $\{\phi_n\}$ is an increasing sequence of step functions such that $\phi_n(y) \to g(y)$ for all y in V outside a null set D. Let
$$S = \{x \in U : h(x) \neq 0\}$$
and
$$C = T^{-1}D \cap S.$$
Then
$$\phi_n(Tx) h(x) \to g(Tx) h(x)$$
for all x in $U \setminus C$, either because $\phi_n(Tx) \to g(Tx)$ ($Tx \notin D$), or because $h(x) = 0$. As D is null, there is an increasing sequence $\{\psi_n\}$ of positive step functions diverging on D and such that $\int \psi_n$ is bounded. Thus
$$\int_U \psi_n(Tx) h(x) \, dx$$
is bounded and
$$\psi_n(Tx) h(x) \to \infty$$
for all x in C, whence C is null. This shows that
$$\phi_n(Tx) h(x) \to g(Tx) h(x)$$
almost everywhere in U, and equation (3) now follows by the Monotone Convergence Theorem.

The extension to functions g in $L^1(V)$ is immediate by linearity. This completes the first part of the proof; the rest should probably be missed out at a first reading. In the sequel we shall give a rather simpler proof of Theorem 1 when we discuss Borel sets.

To simplify the proof of the second half of the theorem we begin by considering the special case where V has *finite* measure. (We have already shown that V is a measurable set.) This implies that $h \in L^1(U)$ and
$$m(V) = \int_U h$$
by equation (3) with $g = \chi_V$.

The next step is to prove an obvious-looking result (which, of course, is a special case of (2)):

TA is null for any null set A contained in U. (∗)

Let A be a null subset of U. For each integer $n \geq 1$, we may cover A by a sequence of bounded open intervals whose total measure is less than $1/n$. Let A_n be the union of this sequence and let
$$G_n = A_1 \cap \ldots \cap A_n.$$
Then $\{G_n\}$ is a decreasing sequence of open sets each of which contains the set A and $m(G_n) < 1/n$. Let G be the intersection of all the sets

G_n; then $A \subset G$ and G is null (because $m(G) \leq m(G_n)$ for all n). As $m(V)$ is finite, TG_n has finite measure

$$m(TG_n) = \int_{G_n} h$$

for all n, and so, by the Monotone Convergence Theorem,

$$m(TG) = \int_G h$$

which is zero as G is null. It follows that TA is null because $TA \subset TG$.

Let us now consider a function $f: U \to \mathbf{R}$ and assume that $fh \in L^1(U)$. Our task is to show that the function $g: V \to \mathbf{R}$ defined by

$$g(Tx) = f(x)$$

is integrable on V and that

$$\int_V g = \int_U fh.$$

From the fact that $fh \in L^1(U)$ we can deduce nothing about the values of f outside

$$S = \{x \in U : h(x) \neq 0\},$$

but we may define $\quad f_0(x) = f(x) \quad$ if $\quad x \in S,$

$\qquad\qquad\qquad\qquad = 0 \qquad$ otherwise.

(In particular $f_0(x) = 0$ if $x \notin U$.) Our assumption that $m(V)$ is finite implies that $h \in L^1(U)$ and so h is certainly measurable. As

$$f_0 = (fh)\,1/h,$$

f_0 is also measurable (Ex. 6.2.5). The equation

$$g_0(Tx) = f_0(x)$$

defines a function $g_0: V \to \mathbf{R}$. Let I be a bounded interval with $\bar{I} \subset U$, K a strictly positive constant and define

$$\bar{f}_0 = \mathrm{mid}\{-K\chi_I, f_0, K\chi_I\},$$

$$\bar{g}_0 = \mathrm{mid}\{-K\chi_{TI}, g_0, K\chi_{TI}\}.$$

As f_0 is measurable, $\bar{f}_0$ is integrable, so there is a sequence $\{\phi_n\}$ of step functions converging to $\bar{f}_0$ almost everywhere in U and we may truncate these ϕ_n's, if necessary, to ensure that they are dominated by $K\chi_I$. The corresponding distorted step functions ψ_n (which vanish outside V) are defined by

$$\psi_n(Tx) = \phi_n(x)$$

and converge to $\bar{g}_0$ *almost everywhere* in V. (This is where we use the crucial statement (∗).) The Dominated Convergence Theorem applied to the equation
$$\int \psi_n = \int \phi_n h$$
now gives
$$\int \bar{g}_0 = \int \bar{f}_0 h.$$

If U is expressed as the union of a sequence of disjoint bounded intervals I with $\bar{I} \subset U$ and then $K \to \infty$ through a sequence of values, the Dominated Convergence Theorem gives
$$\int_V g_0 = \int_U f_0 h.$$

Now $f_0 h = fh$ and so, in the particular case where $f = \chi_U$, we have
$$m(TS) = \int_U h = m(V),$$
so that $g = g_0$ almost everywhere in V and
$$\int_V g = \int_U fh.$$

If E is any measurable set in U then $\chi_E h \in L^1(U)$ and so TE has finite measure
$$m(TE) = \int_E h.$$

This completes the proof when $m(V)$ is finite. In the general case, let $\{J_n\}$ be an increasing sequence of bounded open intervals whose union is $\mathbf{R}^k$, so that V is the union of the bounded sets $J_n \cap V$. The above proof applies to the restriction of T to the open set $T^{-1}J_n$. If we now let n tend to ∞ and apply the Dominated Convergence Theorem, the general result follows.

Corollary. *Let U, V be open subsets of $\mathbf{R}^k$ and T a homeomorphism of U onto V with density function h. Suppose that $h(x) \ne 0$ for almost all x in U and define*
$$g(Tx) = 1/h(x) \quad \text{when} \quad h(x) \ne 0$$
$$= 0 \quad \text{otherwise.}$$
Then g is a density function for T^{-1}.

Proof. Let J be a bounded interval with $\bar{J} \subset V$. According to Proposition 6.3.5, $T^{-1}J$ has finite measure
$$\int_U \chi_{T^{-1}J} = \int_U \chi_{T^{-1}J} \frac{1}{h} h$$

$\left(\text{as } \dfrac{1}{h(x)} h(x) = 1 \text{ for almost all } x \text{ in } U\right)$. Thus, by Theorem 1,

$$m(T^{-1}J) = \int_V \chi_J g = \int_J g.$$

In other words g is a density function for T^{-1}.

The set $S = \{x \in U : h(x) \neq 0\}$ was strategically important in the above proof of Theorem 1. It is not difficult to prove that, for any subset E of U, *TE is measurable if and only if $E \cap S$ is measurable* (see Ex. 5). In practice the domain U of T may often be chosen so that $h \neq 0$ almost everywhere in U, i.e. so that $U \setminus S$ is a null set; in this case TE will be measurable if and only if E is measurable.

The Radon–Nikodym Theorem, which we shall prove in the sequel, gives a necessary and sufficient condition for the existence of a density function h for T, viz. condition (∗) that TA should be null for every null subset A of U. This is a striking result but does not provide any means of calculating h. Our task in the remainder of this section is to introduce a simple-minded kind of derivative $\Delta_T(x)$ and to give sufficient conditions for Δ_T to be a density function for T.

First of all, if h is a density function for T and if p is a point of continuity of h, then, given any $\epsilon > 0$, there is a $\delta > 0$ such that

$$h(p) - \epsilon < h(x) < h(p) + \epsilon$$

for all points x of a cube $C \subset U$ of side less than δ containing p. Thus

$$(h(p) - \epsilon) m(C) \leq \int_C h \leq (h(p) + \epsilon) m(C),$$

i.e.
$$\left| \frac{m(TC)}{m(C)} - h(p) \right| \leq \epsilon$$

for any cube $C \subset U$ of side $< \delta$ containing p (with $m(C) > 0$). We express this condition more succinctly by saying that

$$\frac{m(TC)}{m(C)} \to h(p) \quad \text{as} \quad C \to p.$$

This leads to a general definition. If $p \in U$ and if $m(TC)/m(C)$ converges to a limit $l(p)$ as $C \to p$ then we write

$$l(p) = \Delta_T(p)$$

and call $\Delta_T(p)$ the *measure derivative* of T at p.

We have restricted our definition to cubes rather than general bounded intervals as we want to avoid the embarrassment of 'long thin' intervals. In the definition we have not insisted that C should be open or closed, or even half-open, but in fact we could restrict our attention to any one of these three categories without altering the definition. See Exercise 6. The reader is recommended to consult the masterly treatment of differentiation in [24].

In Chapter 5 we gave a very simple proof of the Fundamental Theorem of the Calculus for a function with *bounded derivative*. The result we shall now prove is the exact analogue in $\mathbf{R}^k$. We need a lemma to take the place of the Mean Value Theorem.

Lemma 1. *If T has measure derivative $\Delta_T(x)$ at every point x of a compact cube $Q \subset U$ and if*
$$\Delta_T(x) \leqslant K$$
for all x in Q, then
$$\frac{m(TC)}{m(C)} \leqslant K$$
for any cube $C \subset Q$ with $m(C) > 0$.

Proof. Assume on the contrary that
$$\frac{m(TC)}{m(C)} = L > K.$$

We may as well assume that C is closed; taking the closure does not alter the denominator and can only increase the numerator. If we now subdivide C into 2^k equal closed subcubes, overlapping only at the faces, then at least one subcube C_1, say, satisfies
$$\frac{m(TC_1)}{m(C_1)} \geqslant L.$$

By continued subdivision we produce a decreasing sequence $\{C_n\}$ of closed cubes which satisfy
$$\frac{m(TC_n)}{m(C_n)} \geqslant L.$$

But these cubes converge to a point x of Q (Ex. 1.3.14) at which we must therefore have
$$\Delta_T(x) \geqslant L.$$

This contradicts the given inequality
$$\Delta_T(x) \leqslant K.$$

Theorem 2. *Let U be an open subset of $\mathbf{R}^k$ and $T: U \to \mathbf{R}^k$ a one–one continuous mapping. If T has measure derivative $\Delta_T(x)$ at every point x of U and if Δ_T is bounded on every compact cube $Q \subset U$, then Δ_T is a density function for T.*

Proof. Cf. Theorem 5.2.3. Let A be a half-open cube whose closure Q is contained in U, and suppose that A is expressed as the union of finitely many disjoint subcubes C. Define the step function

$$\phi = \sum_C \frac{m(TC)}{m(C)} \chi_C$$

whose constant value on C is the 'average density' $m(TC)/m(C)$. Then it is clear that

$$\int \phi = \sum_C m(TC) = m(TA).$$

If we subdivide A successively by halving and use this construction, we produce a sequence $\{\phi_n\}$ of step functions which converges to Δ_T (everywhere) on A and satisfies

$$\int \phi_n = m(TA)$$

for all n. In view of Lemma 1 the Theorem of Bounded Convergence may be applied and gives

$$m(TA) = \int_A \Delta_T.$$

In the first part of the proof of Theorem 1 we only used the density function for half-open cubes, and so this equation extends to sets E for which TE has finite measure and certainly therefore to bounded intervals I with $\bar{I} \subset U$. In other words Δ_T is a density function for T.

Corollary. *If Δ_T is continuous on U, then Δ_T is a density function for T.*

Proof. The continuous real valued function Δ_T is bounded on the compact set Q by Theorem 9 of the Appendix.

This is all very well, but we are not brought up to calculate these measure derivatives! Can we not relate them to something more familiar? It is clear that an affine mapping T has constant measure derivative Δ_T which can be calculated from a square matrix by means of elementary row and column operations as in § 6.3 or, if you like, as the absolute value of a determinant. Our final result shows how to calculate the measure derivative $\Delta_T(x)$ at any point x where T is approximately affine.

We introduce a second *norm* in $\mathbf{R}^k$ by defining

$$\|x\| = \max\{|x_1|, \ldots, |x_k|\}.$$

It is clear that $\qquad \|x - z\| \leq \|x - y\| + \|y - z\|$

and so we may define a new *distance* $\|x-y\|$ between the points x, y. This distance is particularly well suited to our present context as the 'open ball' $\|x-a\| < r$ is an open cube of side $2r$ and the 'closed ball' $\|x-a\| \leq r$ is a closed cube of side $2r$. It can be verified at once that

$$\|x\| \leq |x| \leq \sqrt{k}\|x\|$$

so the notion of open set in $\mathbf{R}^k$, and all the consequent topological notions, are the same for both norms.

We shall say that T is *differentiable* at the point x of U if there is a linear mapping $L_x: \mathbf{R}^k \to \mathbf{R}^k$ such that

$$T(x+h) = T(x) + L_x(h) + o(h) \tag{4}$$

where $o(h)$ tends to 0 more quickly than h. A little more precisely, given any $\epsilon > 0$; there exists $\delta > 0$ such that $x + h \in U$ and

$$\|T(x+h) - T(x) - L_x(h)\| \leq \epsilon \|h\| \tag{5}$$

whenever $\|h\| < \delta$. In this case we refer to L_x as the (*linear*) *derivative* of T at x. We could say that $T(x+h)$ has the 'affine approximation' $T(x) + L_x(h)$ in the neighbourhood of x.

First of all we must show that the linear derivative L_x is *unique* (assuming that it exists at all). One simple way is to recall the *directional derivative*

$$\lim_{t \to 0} (T(x + tu) - T(x))/t$$

for any fixed $u \neq 0$ in $\mathbf{R}^k$. By (4),

$$(T(x+tu) - T(x))/t = L_x(u) + o(t)/t,$$

and this converges to $L_x(u)$ as $t \to 0$. This proves the uniqueness.

If $u_j = (0, 0, \ldots, 1, 0, \ldots, 0)$ with the 1 in the j-th place, the directional derivative $L_x(u_j)$ may be written as

$$\left(\frac{\partial T_1}{\partial x_j}, \ldots, \frac{\partial T_k}{\partial x_j} \right),$$

where $T_1, \ldots, T_k: U \to \mathbf{R}$ are the component functions of T defined by

$$T(x) = (T_1(x), \ldots, T_k(x))$$

and

$$\frac{\partial T_i}{\partial x_j} = \lim_{t \to 0} \{T_i(x_1, \ldots, x_j + t, \ldots, x_k) - T_i(x_1, \ldots, x_j, \ldots, x_k)\}/t$$

is the familiar *partial derivative*. The $(k \times k)$ matrix $(\partial T_i/\partial x_j)$ is usually called the *Jacobian matrix* of T at x; it is the matrix of L_x (referred to the standard basis $\{u_1, \ldots, u_k\}$).

The existence of the partial derivatives, at a point x, follows from the differentiability at x, but the converse is not true (Ex. 9). The simplest *sufficient* condition for differentiability at x is that the partial derivatives $\partial T_i/\partial x_j$ should be continuous at x. Details are given in Exx. 10, 11. We shall say that T is *continuously differentiable on* U if the partial derivatives $\partial T_i/\partial x_j$ are continuous on U.

Recall from p. 142 that any linear mapping $L: \mathbf{R}^k \to \mathbf{R}^k$ satisfies

$$\|L(x)\| \leqslant K\|x\| \quad (x \in \mathbf{R}^k)$$

for some constant K (with due regard to the new norm): see also Ex. 7. It follows at once from (5) that if P, Q are such linear mappings and $Qy = x$, then the linear derivative of PTQ at y is PL_xQ. This observation allows us to simplify the following argument.

Proposition 1. *Let U, V be open subsets of $\mathbf{R}^k$ and T a homeomorphism of U onto V. If T has linear derivative L_x at the point x of U, then T has measure derivative*
$$\Delta_T(x) = \Delta_{L_x}.$$

This is a 'local' result in the sense that it only involves points near x, but it is important for the proof that there should be an open set U containing x and that T should be a homeomorphism of U onto an open subset of $\mathbf{R}^k$.

Proof. As Lebesgue measure is invariant under translations we may simplify the notation by assuming that $x = 0$ and $Tx = 0$; we also write L in place of L_x. Inequality (5) now reads

$$\|T(h) - L(h)\| \leqslant \epsilon \|h\| \tag{6}$$

whenever $\|h\| < \delta$. As in Lemma 6.3.1 we may find linear mappings P, Q with $\Delta_P = \Delta_Q = 1$ so that PLQ has diagonal matrix

$$\begin{pmatrix} d_1 & & & \\ & d_2 & & \\ & & \ddots & \\ & & & d_k \end{pmatrix}.$$

In view of the property
$$\Delta_{PLQ} = \Delta_P \Delta_L \Delta_Q$$

and the remarks just before Proposition 1, we may as well assume that L has diagonal matrix. Thus LC is an *interval* for any cube C.

Case 1. If $\Delta_L = 0$ then at least one of the diagonal elements $d_i = 0$ and the intervals LC lie in the hyperplane with equation $x_i = 0$

(which, of course, is a null set). Given $\epsilon > 0$; we find $\delta > 0$ such that
$$\|T(h) - L(h)\| \leq \epsilon \|h\|$$
for any point h in a cube C containing 0 and of side $s < \delta$. Then TC lies in an interval whose sides are $(|d_i| + 2\epsilon)s$. At least one of these $d_i = 0$ so
$$\frac{m(TC)}{m(C)} \leq (d + 2\epsilon)^{k-1} 2\epsilon,$$
where $d = \max\{|d_1|, ..., |d_k|\}$. Thus
$$\frac{m(TC)}{m(C)} \to 0 \quad \text{as} \quad C \to 0.$$

Case 2. If $\Delta_L \neq 0$ then L has inverse L^{-1}. This allows us to consider $L^{-1}T$ in place of T and so to assume that L is the identity mapping. Thus (6) now reads
$$\|T(h) - h\| \leq \epsilon \|h\| \tag{7}$$
whenever $\|h\| < \delta$. We may assume that $\epsilon < \tfrac{1}{4}$ and choose δ small enough to ensure that the cube $\|x\| < 2\delta$ is contained in U. Let C be a cube containing 0 with side $s < \delta$ and C_-, C_+ the concentric cubes of sides $(1 - 3\epsilon)s, (1 + 3\epsilon)s$, respectively. As $\epsilon < \tfrac{1}{4}$ it is clear that $C_+ \subset U$.

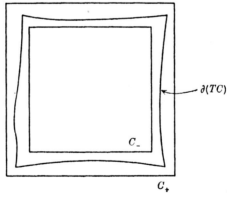

Fig. 48

Inequality (7) shows at once that $TC \subset C_+$. We now apppeal to some simple topology. As T is a homeomorphism of U onto V the boundary ∂C of C carries over to the boundary $\partial(TC)$ of TC. But (7) then shows that $\partial(TC)$ has no points in common with C_-. On the other hand T moves the centre of C a distance of not more than $\tfrac{1}{2}\epsilon s$ and this is less than $\tfrac{1}{2}(1 - 3\epsilon)s$ as $\epsilon < \tfrac{1}{4}$. Thus TC and C_- have at least one point in

common. As C_- is connected we appeal to Theorem 5 of the Appendix to see that $C_- \subset TC$. Now we have the inequalities we have been looking for:
$$C_- \subset TC \subset C_+ \qquad \text{(Fig. 48)}$$
which implies that
$$(1-3\epsilon)^k \leqslant \frac{m(TC)}{m(C)} \leqslant (1+3\epsilon)^k$$
whenever side $C < \delta$. Thus
$$\frac{m(TC)}{m(C)} \to 1 \quad \text{as} \quad C \to 0$$
which gives the required result. (Recall that L is the identity mapping by our construction.)

We may now gather our results together in a final theorem.

Theorem 3. *Let U, V be open subsets of $\mathbf{R}^k$ and T a homeomorphism of U onto V. Let T have (linear) derivative L_x at each point x of U and write $\Delta_T(x) = \Delta_{L_x}$. If Δ_T is bounded on compact subsets of U then Δ_T is a density function for T and*
$$\int_V g(y)\,dy = \int_U g(Tx)\,\Delta_T(x)\,dx. \qquad (8)$$
(If one of these integrals exists, so does the other, and they are equal.)

Formula (8) is often called the *Jacobian* formula for the transformation of integrals; the derivative $\Delta_T(x)$ is the absolute value of the determinant of the Jacobian matrix $(\partial T_i/\partial x_j)$.

In our basic Theorem 1 we assumed that U is an *open* subset of $\mathbf{R}^k$ and T is a *one–one continuous* mapping of U into $\mathbf{R}^k$. There is a famous theorem of Brouwer which states that such a mapping is a *homeomorphism* and the image TU is an *open* subset of $\mathbf{R}^k$ (cf. [18], p. 137). The assumptions in Theorem 3 can therefore be weakened. As long as we insist that T is one–one we may drop the assumption that V is open and that T is a homeomorphism. (The continuity of T follows, of course, from the differentiability. See Ex. 8.) The proof of Brouwer's Theorem requires far more sophisticated topology than we have at our disposal and we do not wish to base any of our results upon it. In practice the continuity of T^{-1} is often harder to check than the other details – the polar transformation is a good case in point, but we could hardly use Brouwer's Theorem to save us this elementary chore. There is another well known result which is treated in most courses on advanced calculus. This uses the slightly stronger condition of continuous differentiability.

6.4] TRANSFORMATIONS 159

The Inverse Mapping Theorem. *Let U be an open subset of $\mathbf{R}^k$ and suppose that $T: U \to \mathbf{R}^k$ is continuously differentiable on U. Then any point x of U at which the linear derivative L_x is invertible, is contained in an open subset of U which is mapped homeomorphically by T onto an open subset of $\mathbf{R}^k$.*

This is just the kind of 'local' condition we need in the proof of Proposition 1, case 2. If we are willing to accept this theorem and the consequent continuity of the derivative Δ_T (Ex. 13) then we have the following very practical result.

Theorem 4. *Let U be an open subset of $\mathbf{R}^k$ and $T: U \to \mathbf{R}^k$ a one–one continuously differentiable mapping. Then $\Delta_T(x) = \Delta_{L_x}$ defines a density function Δ_T for T and*

$$\int_{TU} g(y)\,dy = \int_U g(Tx)\,\Delta_T(x)\,dx.$$

(If one of these integrals exists, then both exist, and they are equal.)

Exercises

1. The *polar transformation* T is defined by $T(r, \theta) = (x, y)$ where $x = r\cos\theta$, $y = r\sin\theta$ ($0 < r, 0 < \theta < 2\pi$). Using the formula for the area of an open or closed circular disk (§4.3, p. 89 and Exx. 4.2.6–8) show that $h(r, \theta) = r$ defines a density function h for T.

2. The *cylindrical polar transformation* T is defined by $T(r, \theta, z) = (x, y, z)$, where $x = r\cos\theta$, $y = r\sin\theta$ ($0 < r$, $0 < \theta < 2\pi$) (Fig. 49).

Use Ex. 1 to show that $h(r, \theta, z) = r$ defines a density function h for T.

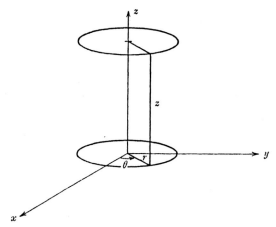

Fig. 49

3. The *spherical polar transformation* T is defined by $T(r, \theta, \phi) = (x, y, z)$ where
$$x = r\sin\theta\cos\phi, y = r\sin\theta\sin\phi, z = r\cos\theta \quad (0 < r, 0 < \theta < \pi, 0 < \phi < 2\pi)$$
(Fig. 50).

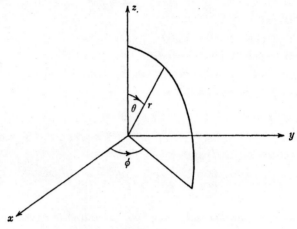

Fig. 50

Let I be the interval in $\mathbf{R}^3$ defined by $0 < r < a$, $\alpha < \theta < \tfrac{1}{2}\pi$, $0 < \phi < 2\pi$ ($0 < \alpha < \tfrac{1}{2}\pi$). Show that the volume of TI is $\tfrac{2}{3}\pi a^3 \cos\alpha$ and hence show that $h(r, \theta, \phi) = r^2 \sin\theta$ defines a density function h for T.

4. Use the polar transformation to evaluate
$$\int_I e^{-(x^2+y^2)} d(x, y),$$
where I is the 'quadrant' $x \geqslant 0$, $y \geqslant 0$, and deduce that
$$\int_0^\infty e^{-x^2} dx = \frac{\sqrt{\pi}}{2}.$$

5. Let $S = \{x \in U : h(x) \neq 0\}$ as defined in the proof of Theorem 1. Show that, for any subset E of U, TE is measurable if and only if $E \cap S$ is measurable.
[Hint: consider first the case where V has finite measure.]

6. Show that in the definition of the measure derivative $\Delta_T(p)$ given on p. 152 we may restrict C to be (i) an open cube containing p, or (ii) a half-open cube, of type $[\,,\,)^k$, containing p, or (iii) a closed cube, of strictly positive measure, containing p.

7. The linear mapping $L : \mathbf{R}^k \to \mathbf{R}^k$ has matrix (c_{ij}). Show that
$$\|L(x)\| \leqslant K\|x\|,$$
where $K = \max\{|c_{i1}| + \ldots + |c_{ik}| : 1 \leqslant i \leqslant k\}$.

6.4] TRANSFORMATIONS

8. If T is differentiable at x, as defined on p. 155, show that T is continuous at x.

9. Let
$$f(x_1, x_2) = \frac{x_1 x_2^2}{x_1^2 + x_2^4} \quad \text{if} \quad (x_1, x_2) \neq (0, 0) \quad \text{and} \quad f(0, 0) = 0.$$
Show that the directional derivative.
$$\lim_{t \to 0} \{(f(x + tu) - f(x))/t\}$$
exists for each fixed $u \neq 0$ in $\mathbf{R}^2$ but that f is not continuous at 0.

10. Let U be an open subset of $\mathbf{R}^2$, $f: U \to \mathbf{R}$, and define
$$D_1 f(x) = \lim_{t \to 0} \{(f(x_1 + t, x_2) - f(x_1, x_2))/t\},$$
$$D_2 f(x) = \lim_{t \to 0} \{(f(x_1, x_2 + t) - f(x_1, x_2))/t\}.$$
(This is more a precise notation than the classical $\partial f/\partial x_1$, $\partial f/\partial x_2$.) Assume that $D_1 f, D_2 f$ (exist near x and) are continuous at x. Apply the Mean Value Theorem to
$$f(x_1 + h_1, x_2 + h_2) - f(x_1, x_2 + h_2), \quad f(x_1, x_2 + h_2) - f(x_1, x_2),$$
and show that $\quad f(x + h) = f(x) + h_1 D_1 f(x) + h_2 D_2 f(x) + o(h)$
where $\quad o(h)/\|h\| \to 0 \quad \text{as} \quad \|h\| \to 0$.

Suppose now that $T: U \to \mathbf{R}^2$ and the partial derivatives $D_j T_i$ are continuous on U. Apply the above argument to T_1, T_2 and deduce that T is differentiable on U.

11. Let U be an open subset of $\mathbf{R}^k$. Extend Ex. 10 to a mapping
$$T: U \to \mathbf{R}^k.$$

12. Verify that the three polar transformations of Exx. 1, 2, 3 are continuously differentiable. Find the Jacobian matrix in each case and calculate their density functions.

13. Let T be continuously differentiable on U. Show that Δ_T is continuous on U, (i) by assuming that Δ_T is the absolute value of a determinant; (ii) by adapting the proof of Proposition 1.

14. Let $A = \{(x, y): x > 0, y > 0, 0 < xy < 3, x < y < 2x\}$. Use the substitution $s = xy$, $t = y/x$ to evaluate $\int_A y^2 d(x, y)$.

7
THE SPACES L^p

From the beginning we have stressed the importance of the completeness of the real numbers. To round off this volume we return to the question of completeness in the far more sophisticated setting of the spaces L^p ($p \geq 1$): these consist of the measurable functions $f\colon \mathbf{R}^k \to \mathbf{R}$ for which $|f|^p$ is integrable (over some given subset of $\mathbf{R}^k$). In §7.1 the completeness of $\mathbf{R}$ is re-examined in terms of distance (rather than order), and the completeness of L^p is proved in §7.2. In §7.3 and §7.4 we stress the geometric nature of the space L^2 which has many of the properties we associate with Euclidean space $\mathbf{R}^k$, including a notion of duality. These geometric methods are remarkably powerful (see especially the Exx. 7.4.5–10 associated with Clarkson's Inequalities and the extension of the idea of duality to L^p, L^q); in §7.5 they give further insight into the expansions in terms of orthogonal functions, and guarantee the (strong) convergence of the Fourier series of *any* function in L^2. The power of the Lebesgue integral is further illustrated in §7.6 where some of the most famous classical theorems on the (pointwise) convergence of Fourier series are proved. The last section is in a lighter vein and points the way to general Banach and Hilbert spaces which are of great importance in the study of modern functional analysis.

7.1 The completeness of R as a metric space

At the end of Chapter 1 we promised to formulate the notion of completeness in terms of *distance* rather than in terms of order. To motivate this let us consider the confidence we have that any decimal expansion is 'convergent'. On what intuitive grounds is this confidence based? The student who is familiar with the material in Chapter 1 may well answer, 'Because the partial sums increase and are bounded above.' But for most people the confidence is based on the fact that any two expansions which agree to, say, 5 decimal places, whatever they do thereafter, cannot differ by more than 10^{-5}. The point is that it is essentially finite sums that are being compared with one another, rather than finite sums with an infinite one. In more formal language, we shall expect a sequence $\{s_n\}$ of real numbers to be convergent if

7.1] COMPLETENESS OF R

the terms are ultimately close to one another. This is made precise by **Cauchy's General Principle of Convergence.**

Theorem 1. *The following condition is necessary and sufficient for the sequence $\{s_n\}$ of real numbers to converge to a real number.*

Given $\epsilon > 0$; there exists an integer N such that
$$|s_m - s_n| < \epsilon \quad \text{for} \quad m, n \geq N.$$

This condition is also written:
$$|s_m - s_n| \to 0 \quad \text{as} \quad m, n \to \infty.$$

Proof. (i) The necesssity is quite obvious, for if all the terms are ultimately near some limit s, then they must be near one another: if
$$|s_n - s| < \epsilon \quad \text{for} \quad n \geq N,$$
then $\quad |s_m - s_n| \leq |s_m - s| + |s_n - s| < 2\epsilon \quad \text{for} \quad m, n \geq N.$

(ii) Suppose that the Cauchy condition is satisfied. We may find a positive integer $N(1)$ such that
$$|s_n - s_{N(1)}| < 2^{-1} \quad \text{for} \quad n \geq N(1),$$
then a positive integer $N(2) > N(1)$ such that
$$|s_n - s_{N(2)}| < 2^{-2} \quad \text{for} \quad n \geq N(2),$$
and so on. By induction we find a subsequence $\{s_{N(k)}\}$ of $\{s_n\}$ such that
$$|s_n - s_{N(k)}| < 2^{-k} \quad \text{for} \quad n \geq N(k)$$
$(k = 1, 2, \ldots)$. By the Comparison Test
$$s_{N(1)} + (s_{N(2)} - s_{N(1)}) + (s_{N(3)} - s_{N(2)}) + \ldots$$
is absolutely convergent, and hence convergent by Proposition 1.1.2. In other words the subsequence $\{s_{N(k)}\}$ converges to a limit s, say. Given any $\epsilon > 0$; we may find k large enough to ensure that
$$|s_{N(k)} - s| < \epsilon$$
and also $\quad 2^{-k} < \epsilon.$

Then $\quad |s_n - s| \leq |s_n - s_{N(k)}| + |s_{N(k)} - s| < 2\epsilon \quad \text{for} \quad n \geq N(k),$
and so $\{s_n\}$ converges to s.

(For an alternative proof of the sufficiency see Ex. 5.2.1.)

Let X be a *metric space*, as defined in the Appendix, i.e. X is a non-empty set on which a distance d is defined. Let $s_n \in X$ for $n = 1, 2, \ldots$;

we shall say that the sequence $\{s_n\}$ *converges in* X if there is a point s in X for which $d(s_n, s) \to 0$ as $n \to \infty$. This agrees with our definition of convergence in $\mathbf{R}$ where the distance d is given, as usual, by

$$d(x, y) = |x - y|.$$

If
$$d(s_m, s_n) \to 0 \quad \text{as} \quad m, n \to \infty$$

then we shall call $\{s_n\}$ a *Cauchy sequence. A complete metric space* is now one in which every Cauchy sequence converges. In view of Theorem 1, $\mathbf{R}$ is a complete metric space with respect to the given distance.

This further use of the word 'complete' in describing the real line $\mathbf{R}$ calls into question the terminology of Chapter 1. The Axiom of Completeness should perhaps have been called the Axiom of Order-Completeness. In the presence of the axioms for an ordered field, as described in §1.3, together with the Archimedean Axiom that the integers are unbounded, it is now quite a simple matter to deduce the Axiom of Order-Completeness from the completeness of $\mathbf{R}$ as a metric space. This is set as Ex. 1. As we are perfectly happy to include the Axiom of Archimedes in our list of assumptions about the real line, we may now use this new description of the completeness of $\mathbf{R}$ with a clear conscience.

Exercise

1. Let $\{s_n\}$ be an increasing sequence of real numbers bounded above by K.

(i) Use Archimedes' Axiom to prove that $2^{-n} \to 0$.

(ii) By successive halving of the interval $[s_1, K]$ find a subsequence of $\{s_n\}$ that satisfies the Cauchy condition.

(iii) Use the completeness of $\mathbf{R}$ as a metric space to prove that $\{s_n\}$ converges to a real number.

7.2 The spaces L^p

Let p be any real number satisfying $p \geq 1$ and denote by L^p the set of all (Lebesgue) measurable functions $f: \mathbf{R} \to \mathbf{R}$ such that $|f|^p$ is (Lebesgue) integrable. In view of Proposition 6.1.1, Corollary, this agrees with our previous use of L^1 to denote the space of integrable functions $f: \mathbf{R} \to \mathbf{R}$. It makes virtually no difference to our discussion if we consider an interval I in $\mathbf{R}^k$ and the space $L^p(I)$ of measurable functions $f: \mathbf{R}^k \to \mathbf{R}$ such that $|f|^p$ is integrable on I. For definiteness we write L^p for $L^p(\mathbf{R})$, but all that follows will be true of $L^p(I)$, provided only that $m(I) > 0$.

If $f \in L^p$ and $a \in \mathbf{R}$, then af is measurable and $|af|^p \in L^1$, so that $af \in L^p$. If $f, g \in L^p$ then

$$|f+g| \leq 2 \max\{|f|, |g|\}$$

and $$|f+g|^p \leq 2^p \max\{|f|^p, |g|^p\};$$

it follows that $f+g \in L^p$. We have therefore shown that L^p *is a linear space.*

For any f in L^p we define the *norm*

$$\|f\| = \left\{\int |f|^p\right\}^{1/p}.$$

From this definition it follows that

N 1. $\|f\| \geq 0$,

N 2. $\|f\| = 0$ if and only if $f = 0$ almost everywhere,

N 3. $\|af\| = |a|\|f\|$,

N 4. $\|f+g\| \leq \|f\| + \|g\|$,

($f, g \in L^p, a \in \mathbf{R}$). As for Euclidean space, N 4 is called the *triangle inequality*; it will be established in Proposition 2. Rules N 1 and N 3 are immediate from the definition of $\|f\|$, and N 2 follows from the Monotone Convergence Theorem, Corollary. The words 'almost everywhere' in N 2 are something of an embarrassment. Strictly speaking, the above norm should be called a *pseudo-norm* (which term we dislike). Alternatively, we may define a new space $\mathscr{L}^p$ whose elements are equivalence classes $\tilde{f}$ of functions in L^p according to the equivalence relation:

$$f \sim g \quad \text{if and only if} \quad f = g \text{ a.e.}$$

and define $$\|\tilde{f}\| = \|g\|,$$

where g is any representative of the equivalence class $\tilde{f}$. It is obvious that this definition does not depend on the choice of representative g. The linear space structure on $\mathscr{L}^p$ is similarly defined in terms of representatives from L^p:

$$\tilde{f} + \tilde{g} = (f+g)^\sim$$
$$a\tilde{f} = (af)^\sim.$$

In this way we get a norm on $\mathscr{L}^p$ which has the desired property that

$$\|\tilde{f}\| = 0 \quad \text{if and only if} \quad \tilde{f} = \tilde{0}.$$

In this context it is convenient to refer to the elements of $\tilde{0}$ as *null functions*.

There is a gentlemen's agreement, to which we subscribe, that if one refers to L^p as a *normed linear space* it is really $\mathscr{L}^p$ that is meant. In the same way, we shall define the distance d by means of
$$d(f,g) = \|f-g\|,$$
and refer to L^p as a metric space, when we really mean that $\mathscr{L}^p$ is a metric space under the definition
$$d(\tilde{f},\tilde{g}) = d(f,g).$$

If $p > 1$ we shall see that the space L^p is paired with the space L^q where
$$\frac{1}{p} + \frac{1}{q} = 1$$
(whence $q > 1$). At a later stage we shall prove that each of these spaces is, in a deeply significant sense, the *dual* of the other. But an essential relation between L^p and L^q is already brought out as we seek to establish the simple triangle inequality.

Lemma 1. *If $a, b \geq 0$ then*
$$ab \leq \frac{a^p}{p} + \frac{b^q}{q}.$$

Proof. Consider the points (x, y) in the first quadrant ($x \geq 0, y \geq 0$) satisfying
$$x^p = y^q,$$
and the areas S_a, S_b shown in Fig. 51.

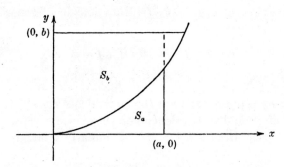

Fig. 51

As $(1/p) + (1/q) = 1$, the relation $x^p = y^q$ may be written as
$$y = x^{p-1}$$
or
$$x = y^{q-1}$$

and so these areas are
$$S_a = \int_0^a x^{p-1}\,dx = \frac{a^p}{p},$$
$$S_b = \int_0^b y^{q-1}\,dy = \frac{b^q}{q}.$$

It is clear from the figure that the sum of these areas is not less than the rectangular area ab. (Moreover, there is equality if and only if $a^p = b^q$.)

The reader may prefer the rather more conventional proof of this inequality in Ex. 2.

Proposition 1 (*Hölder's Inequality*). *Let $p, q > 1$ satisfy the relation $(1/p) + (1/q) = 1$. If $f \in L^p$ and $g \in L^q$ then $fg \in L^1$ and*
$$\int |fg| \leqslant \|f\|_p \|g\|_q.$$

Here we have used the suffixes p, q to distinguish the norms in L^p, L^q, respectively.

Proof. Hölder's Inequality is homogeneous in the sense that if it is satisfied for functions f, g then it will also be satisfied for af, bg ($a, b \in \mathbf{R}$). It is clearly satisfied if f or g is null. Without loss of generality we may therefore assume that
$$\|f\|_p = \|g\|_q = 1,$$
i.e.
$$\int |f|^p = \int |g|^q = 1.$$

According to Lemma 1
$$|fg| \leqslant \frac{|f|^p}{p} + \frac{|g|^q}{q}.$$

Thus the measurable function fg is integrable and
$$\int |fg| \leqslant \frac{1}{p} + \frac{1}{q} = 1.$$

Proposition 2 (*Minkowski's Inequality*). *If $f, g \in L^p$ ($p \geqslant 1$), then*
$$\|f + g\| \leqslant \|f\| + \|g\|.$$

Proof. The case $p = 1$ is already known, so assume that $p > 1$. First of all
$$|f+g|^p \leqslant |f||f+g|^{p-1} + |g||f+g|^{p-1}. \tag{1}$$

As $(p-1)q = p$ it follows that $|f+g|^{p-1} \in L^q$; write A for the norm of $|f+g|^{p-1}$ in L^q, i.e.
$$A = \left\{\int |f+g|^p\right\}^{1/q}.$$
By Hölder's Inequality applied to (1) we have
$$\int |f+g|^p \leq (\|f\| + \|g\|) A. \tag{2}$$
If $A = 0$ then $\|f+g\| = 0$ and the result is obvious; if $A \neq 0$ then we divide both sides of (2) by A and use the relation
$$1 - \frac{1}{q} = \frac{1}{p}$$
to deduce that
$$\left\{\int |f+g|^p\right\}^{1/p} \leq \|f\| + \|g\|,$$
as required.

Before we go on to prove the centrally important theorem that these L^p spaces are complete, it may be as well to note that there are at least three kinds of convergence for a sequence $\{f_n\}$ whose terms are in L^p. If $\{f_n(x)\}$ converges to $f(x)$ (in **R**) for each x (in **R**) then we say that $\{f_n\}$ *converges pointwise* to the function f. In a similar way, if $\{f_n(x)\}$ converges to $f(x)$ for almost all x, we say that $\{f_n\}$ *converges (pointwise) almost everywhere* to the function f. The limit function in either of these cases may, or may not, belong to L^p. According to the definition in §7.1, $\{f_n\}$ *converges to f in L^p* if $f \in L^p$ and $\|f_n - f\| \to 0$ as $n \to \infty$. For emphasis this last kind of convergence is often called *strong convergence* or *convergence in mean (of order p)*. We leave the reader to verify that if $\{f_n\}$ converges strongly to both f and g, then $f = g$ a.e.

Example 1. Let
$$f_n(x) = n \quad \text{for} \quad x \quad \text{in} \quad (0, 1/n),$$
$$= 0 \quad \text{otherwise}.$$
The sequence $\{f_n\}$ converges pointwise (everywhere) to the zero function. On the other hand
$$\int |f_n|^p = n^{p-1}$$
so that
$$\|f_n\| = n^{1-1/p} \to \infty \quad \text{as} \quad n \to \infty$$
for $p > 1$. Even if $p = 1$, the picture is not much better, for although $\|f_n\| = 1$ for all n, there is still no function f in L^1 for which
$$\|f_n - f\| \to 0.$$
The reader is invited to prove this final statement in Ex. 3.

7.2]   THE SPACES L^p

Example 2. The sequence

$$\chi_{[0,1]}, \quad \chi_{[0,1/2]}, \quad \chi_{[1/2,1]}, \quad \chi_{[0,1/4]}, \quad \chi_{[1/4,1/2]}, \quad \cdots$$

of characteristic functions may be written as $\{\chi_{I_n}\}$, where

$$I_n = [n2^{-k} - 1, (n+1)2^{-k} - 1]$$

and k is the unique integer satisfying $2^k \leqslant n < 2^{k+1}$.

We verify in this case that

$$\|f_n\| = 2^{-k/p}$$

and so $\|f_n\| \to 0$ as $n \to \infty$. In other words, $\{f_n\}$ converges in mean to zero. But $\{f_n(x)\}$ does not converge for any x in $[0, 1]$.

These two examples show that, in the case of L^p, pointwise convergence does not imply strong convergence, and strong convergence does not imply pointwise convergence. It is interesting to compare this situation with Euclidean space $\mathbf{R}^k$ in which the element

$$x = (x_1, \ldots, x_k)$$

may be thought of as a real valued function x defined on the set $\{1, \ldots, k\}$. In this case strong convergence and pointwise convergence mean exactly the same thing (see Ex. 4).

As L^p is a linear space we may also interpret convergence of series. Let $\{a_n\}$ be a sequence whose terms are in L^p and define the partial sums

$$s_n = a_1 + a_2 + \ldots + a_n$$

for $n \geqslant 1$. We say that Σa_n *converges to s in L^p* if $s \in L^p$ and $\|s_n - s\| \to 0$ as $n \to \infty$.

As early as §1.1 we saw the relevance of absolute convergence to the completeness of $\mathbf{R}$, and this emerged again in the last section when we proved Cauchy's General Principle of Convergence. The following result is an extension to L^p of the Monotone Convergence Theorem interpreted in terms of series (as in Ex. 5.1.8).

Theorem 1. *Let $a_n \in L^p$ for $n = 1, 2, \ldots$. If $\Sigma \|a_n\|$ converges in $\mathbf{R}$, then Σa_n converges to a function s in L^p. Moreover $\Sigma a_n(x)$ converges to $s(x)$ for almost all x.*

Proof. Let $\Sigma \|a_n\| = M$ and let

$$g_n = |a_1| + \ldots + |a_n|$$

for $n \geqslant 1$. Then $\{g_n\}$ is an increasing sequence of positive functions in L^p and

$$\|g_n\| \leqslant \|a_1\| + \ldots + \|a_n\|$$

by the triangle inequality (extended to n terms). Thus
$$\|g_n\| \leq M$$
i.e.
$$\int g_n^p \leq M^p.$$

By the Monotone Convergence Theorem the increasing sequence $\{g_n^p\}$ converges almost everywhere to a positive function h in L^1 and
$$\int h \leq M^p.$$

Thus $\{g_n\}$ converges almost everywhere to a positive measurable function g, where $g^p = h$, so that $g \in L^p$ and $\|g\| \leq M$.

By Proposition 1.1.2, Σa_n converges almost everywhere to a function s where $|s| \leq g$. Also s is measurable, by Theorem 6.1.1; thus $s \in L^p$, and $\|s\| \leq M$. In terms of the partial sums
$$s_n = a_1 + \ldots + a_n$$
this means that
$$s_n - s \to 0 \text{ a.e.}$$
But
$$|s_n - s| \leq 2g$$
and so by the Dominated Convergence Theorem,
$$\int |s_n - s|^p \to 0.$$
Thus
$$\|s_n - s\| \to 0$$
i.e. Σa_n converges to s in L^p.

As a direct consequence of Theorem 1 we have our main result.

Theorem 2. *The metric space L^p is complete.*

Proof. Let $\{f_n\}$ be a Cauchy sequence in L^p. As in the proof of Theorem 7.1.1, we find positive integers
$$N(1) < N(2) < \ldots$$
so that
$$\|f_n - f_{N(k)}\| < 2^{-k} \quad \text{for} \quad n \geq N(k)$$
($k = 1, 2, \ldots$). Then the series
$$\|f_{N(1)}\| + \|f_{N(2)} - f_{N(1)}\| + \|f_{N(3)} - f_{N(2)}\| + \ldots$$
is convergent by the Comparison Test. Now by Theorem 1, the series
$$f_{N(1)} + (f_{N(2)} - f_{N(1)}) + (f_{N(3)} - f_{N(2)}) + \ldots$$

converges to a function f in L^p. In other words,
$$\|f_{N(k)} - f\| \to 0 \quad \text{as} \quad k \to \infty.$$
Given any $\epsilon > 0$, we find k so that
$$\|f_{N(k)} - f\| < \epsilon$$
and also $2^{-k} < \epsilon$.

Then $$\|f_n - f\| \leq \|f_n - f_{N(k)}\| + \|f_{N(k)} - f\| < 2\epsilon$$
for $n \geq N(k)$, and so $\{f_n\}$ converges to f in L^p.

The above proof of Theorem 2 in fact gives rather more:

Theorem 3. *Let $\{f_n\}$ be a Cauchy sequence in L^p. Then there is a subsequence of $\{f_n\}$ which converges pointwise almost everywhere to a function f, where $f \in L^p$ and $\{f_n\}$ converges strongly to f.*

Needless to say, if $\{f_n\}$ also converges strongly to g, then $f = g$ a.e. In other words, the equivalence class $\tilde{f}$ in $\mathscr{L}^p$ is determined uniquely by the Cauchy sequence $\{f_n\}$.

It is instructive to look again at Example 2. The sequence $\{f_n\}$ satisfies the Cauchy condition
$$\|f_m - f_n\| \to 0 \quad \text{as} \quad m, n \to \infty$$
(because $\{f_n\}$ converges strongly), but $\{f_n(x)\}$ fails to converge for all points x in $[0, 1]$. Thus it is essential in Theorem 3 to select a suitable subsequence which converges pointwise almost everywhere to the function f. In Example 2 the subsequence
$$\chi_{[0,1]}, \quad \chi_{[0,1/2]}, \quad \chi_{[0,1/4]}, \ldots$$
converges to zero almost everywhere.

If it happens that the sequence $\{f_n\}$ satisfies a stronger condition of the Cauchy type such as
$$\|f_{n+1} - f_n\| < 2^{-n}$$
for $n \geq 1$ – we might call this a *rapid Cauchy sequence* – then we may apply Theorem 1 directly to the series
$$f_1 + (f_2 - f_1) + (f_3 - f_2) + \ldots.$$
It follows in this case that $\{f_n\}$ converges almost everywhere to f in L^p. Of course, the number 2 is not particularly important in the above discussion; all we have used is that $\Sigma 2^{-n}$ is convergent.

Exercises

1. Show that the null functions form a linear subspace of L^p. Show that the definitions $\tilde{f}+\tilde{g} = (f+g)^\sim, a\tilde{f} = (af)^\sim$ for the addition and multiplication by scalars in $\mathscr{L}^p$ are unambiguous.

2. Let c be a real number satisfying $0 < c < 1$ and define
$$f(x) = (1-c) + cx - x^c$$
for $x \geqslant 0$. By differentiating f and applying the Mean Value Theorem, show that $f(x) \geqslant 0$ (with equality only for $x = 1$). Deduce the inequality of Lemma 1 (when $b \neq 0$) by taking $c = 1/p$, $x = a^p/b^q$.

3. If $p = 1$, show that the sequence $\{f_n\}$ of Example 1 does not converge strongly to any function f in L^1.

4. Show that strong and pointwise convergence are equivalent in $\mathbf{R}^k$.

5. Suppose that $0 < s \leqslant 1$ and let L^s consist of all measurable functions f for which $|f|^s \in L^1$. Show that L^s is a linear space. If $\|f\| = (\int |f|^s)^{1/s}$ show that
$$\|f+g\| \geqslant \|f\| + \|g\|$$
for any *positive* f, g in L^s.
(Note that the sense of Minkowski's Inequality has been reversed.)

7.3 The geometry of L^2

If $f, g \in L^2$ and $a, b \in \mathbf{R}$, then
$$(af+bg)^2 = a^2 f^2 + 2ab fg + b^2 g^2;$$
as $f^2, g^2 \in L^1$ and the measurable function fg satisfies
$$|fg| \leqslant \tfrac{1}{2}(f^2 + g^2),$$
it follows from Proposition 6.1.1 that $fg \in L^1$ and so $(af+bg)^2 \in L^1$. Thus the measurable function $af+bg \in L^2$. This establishes, once again, the fact that L^2 is a linear space over $\mathbf{R}$.

For any f, g in L^2 we have just seen that $fg \in L^1$; let us define the *scalar product* (or *inner product*)
$$(f, g) = \int fg$$
of any pair of functions f, g in L^2. For future reference we list some of the properties of this scalar product. For f, g, h in L^2 and a, b in $\mathbf{R}$,

SP 1. $(f, g) = (g, f)$,
SP 2. $(af+bg, h) = a(f, h) + b(g, h)$,
SP 3. $(f, f) \geqslant 0$,
SP 4. $(f, f) = 0$ if and only if $f = 0$ a.e.

The first three of these are immediately verified and SP 4 follows from the Monotone Convergence Theorem, Corollary. We may define the *norm* in terms of the scalar product, viz.

$$\|f\| = \sqrt{(f,f)} \quad (f \in L^2).$$

The Hölder and Minkowski inequalities are so much easier in this case that it is worth deriving them independently; in so doing we only use rules SP 1, SP 2, SP 3 above.

Proposition 1 (Schwarz' Inequality).

$$|(f,g)| \leq \|f\| \|g\| \quad (f, g \in L^2).$$

Proof. Exactly as for Proposition 6.3.3.

Proposition 2 (The Triangle Inequality).

$$\|f+g\| \leq \|f\| + \|g\| \quad (f, g \in L^2).$$

Proof. Exactly as for Proposition 6.3.4.

If we define the distance d by the equation

$$d(f,g) = \|f-g\|$$

then the triangle inequality takes on the more familiar look:

$$d(f,g) \leq d(f,h) + d(h,g) \quad (f, g, h \in L^2)$$

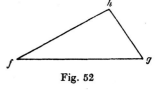

Fig. 52

(see Fig. 52). This says that, for any triangle in L^2, the length of any one side is no greater than the sum of the lengths of the other two sides.

Suppose that $\|f\|$, $\|g\|$ are both different from zero. If one cares to quote from analysis that there is a continuous one-one function *cos* which carries the interval $[0, \pi]$ onto the interval $[-1, 1]$, and vanishes at $\pi/2$ (Fig. 53), then it is possible, in the light of Schwarz' Inequality, to define a unique real number θ in $[0, \pi]$ which satisfies

$$(f, g) = \|f\| \|g\| \cos \theta. \tag{1}$$

Fig. 53

The number θ is the *angle between f and g* (Fig. 54). Formula (1) for the scalar product is almost certainly familiar from elementary vector analysis, though the assumptions there were probably different! In fact we shall not use this concept of angle, but we shall make a great deal of the closely related notion of orthogonality, which corresponds to the case where $\theta = \pi/2$. Let us drop the assumption that f and g are non-zero and say that f, g are *orthogonal* if $(f, g) = 0$. An encouraging start is the following famous result.

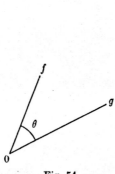
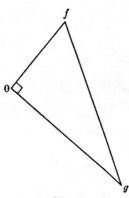

Fig. 54   Fig. 55

Proposition 3 (Pythagoras' Theorem). *If f, g are orthogonal, then*

$$\|f \pm g\|^2 = \|f\|^2 + \|g\|^2.$$

In the first place this proposition relates to a triangle which is right angled at 0 (Fig. 55), but the distance, given in terms of the norm, is clearly invariant under translation:

$$d(f+h, g+h) = d(f, g) \quad (f, g, h \in L^2)$$

and so the result applies to any right angled triangle in L^2.

Proof. By rules SP 1, SP 2, SP 3,

$$(f \pm g, f \pm g) = (f, f) \pm 2(f, g) + (g, g)$$

and we are given that $(f, g) = 0$.

In any linear space we may refer to the set of points

$$\{(1-t)f + tg : 0 \leqslant t \leqslant 1\}$$

as the *(straight line) segment* joining f to g. This notion is greatly strengthened if we have a distance defined in terms of a norm which

satisfies the rules N 1 to N 4 of the previous section. In particular, the rule
$$\|af\| = |a|\,\|f\|$$
shows that distance is related linearly to multiplication by scalars, and so the point
$$(1-t)f+tg$$
is identified as the one which *divides f, g in the ratio $t: 1-t$* (Ex. 2; see Fig. 56).

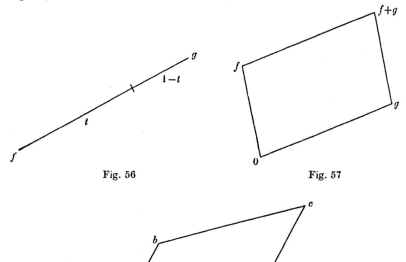

Fig. 56 Fig. 57

Fig. 58

In particular $\tfrac{1}{2}(f+g)$
is the *mid point* of f and g. The segment joining 0 to f translates to the segment joining g to $f+g$; we shall say that the four points $0, f, f+g, g$ are the vertices of a *parallelogram* in L^2 (Fig. 57). More generally we may say that a, b, c, d are the vertices of a *parallelogram* if and only if $a+c = b+d$ (Fig. 58).

Proposition 4 (*The Parallelogram Law*).

$$\|f+g\|^2 + \|f-g\|^2 = 2\|f\|^2 + 2\|g\|^2 \quad (f, g \in L^2).$$

This may be interpreted as saying that, for any parallelogram in L^2, the sum of the squares of the lengths of the diagonals is equal to the sum of the squares of the lengths of the sides.

Proof. Expand $(f+g, f+g), (f-g, f-g)$ using rules SP 1 to SP 3 and add.

A subset S of L^2 is said to be *convex* if S is non-empty and, for every pair of points f, g in S, the whole segment joining f to g also lies in S

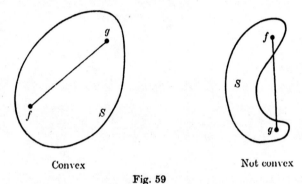

Convex   Not convex

Fig. 59

(Fig. 59). Among the most important convex subsets of L^2 are the linear subspaces: a non-empty subset M of L^2 is a *linear subspace* of L^2 if
$$f, g \in M \Rightarrow af + bg \in M$$
for all a, b in $\mathbf{R}$. If the context makes it clear that linear spaces are under discussion, then a linear subspace is usually called a subspace, for short.

We now come to a powerful result which makes full use of the parallelogram law – this law is not available for any of the other L^p spaces (see Ex. 3) but the argument may be modified to prove Theorem 1 for any L^p with $p > 1$ (see Exx. 4–6). The main idea of the proof may be expressed vividly by imagining a parallelogram with equal sides of fixed length freely hinged at the corners, being stretched along one diagonal by pulling two opposite vertices apart: the effect is to force the other two vertices together (see Fig. 60). Anyone who has put up a garden trellis will know what this means in practice!

Theorem 1. *If S is a closed convex subset of L^2 and f any point of L^2, then there is a unique point g of S which is nearest to f* (Fig. 61).

The word 'closed' here refers to the topology on L^2 regarded as a metric space. As we should expect, uniqueness is strictly speaking in $\mathscr{L}^2$: thus g is uniquely determined to within a null function.

7.3] GEOMETRY OF L^2 177

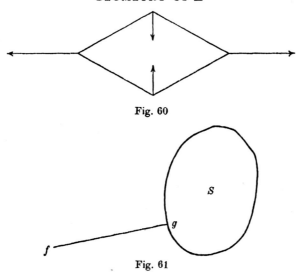

Fig. 60

Fig. 61

Proof. The translate $\quad S-f = \{s-f : s \in S\}$

is a closed convex subset of L^2 and so we may assume without loss of generality that $f = 0$. Let $d = \inf\{\|s\| : s \in S\}$.

Consider first the uniqueness. Suppose that

$$g, h \in S \quad \text{and} \quad \|g\| = \|h\| = d.$$

As S is convex the mid point $\tfrac{1}{2}(g+h)$ also belongs to S. Thus

$$\|\tfrac{1}{2}(g+h)\| \geq d$$

or equivalently, $\quad \|g+h\| \geq 2d.$

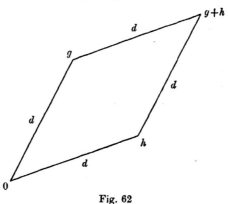

Fig. 62

In Fig. 62 the vertices $0, g+h$ have been forced their maximum distance apart and so g, h must coincide. Formally, we apply the parallelogram law to see that

$$\|g-h\|^2 = 2\|g\|^2 + 2\|h\|^2 - \|g+h\|^2 \leq 0$$

and so $\|g-h\| = 0$. This gives $g = h$ almost everywhere.

The existence of g is proved in a similar way, but uses the fact that S is closed and L^2 is complete. By the definition of d there is a sequence $\{s_n\}$ in S for which
$$\|s_n\| \to d.$$

By the parallelogram law, as above,

$$\|s_m - s_n\|^2 \leq 2\|s_m\|^2 + 2\|s_n\|^2 - 4d^2$$

and so $\|s_m - s_n\| \to 0$ as $m, n \to \infty$.

In other words $\{s_n\}$ is a Cauchy sequence, and so, by Theorem 7.2.2, $\{s_n\}$ converges to a point g in L^2. As $s_n \in S$ for all n and S is closed, it follows that $g \in S$ (Appendix p. 227). Finally, let n tend to infinity in the inequality
$$\|g\| \leq \|s_n\| + \|g - s_n\|$$

and deduce that
$$\|g\| \leq d$$

from which the result follows.

If S is a straight line and $f \notin S$, then our intuition, based on Euclidean experience, suggests that the nearest point g will be obtained by 'dropping a perpendicular'. This generalises to L^2.

Theorem 2. *Let M be a closed linear subspace of L^2, f a point of L^2 outside M and g the unique point of M nearest to f (Fig. 63). Then $f - g$ is orthogonal to every element of M.*

The result is still true if $f \in M$, but in this case, of course,

$$f = g \text{ (a.e.)}.$$

Proof. Let h be an arbitrary element of M. Then, by the definition of g,
$$\|f - g\|^2 \leq \|f - g - th\|^2$$
for all t in **R**. Thus

$$\|f - g\|^2 \leq \|f - g\|^2 - 2t(f - g, h) + t^2\|h\|^2$$

or
$$0 \leq -2t(f - g, h) + t^2\|h\|^2 \tag{2}$$

for all t in **R**. From this it follows that

$$(f - g, h) = 0,$$

otherwise we could choose t near 0 to contradict (2).

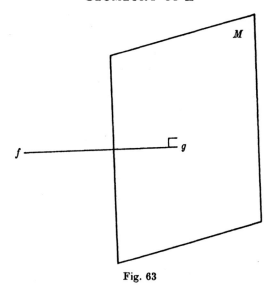

Fig. 63

If S is any subset of L^2 then we denote by $S^\perp$ the set of all elements that are orthogonal to S, i.e.

$$S^\perp = \{f \in L^2 : (s, f) = 0 \quad \text{for all } s \text{ in } S\}.$$

If S consists of the single element s, we shall write $s^\perp$ in place of $\{s\}^\perp$. There is no difficulty in showing that $S^\perp$ is a linear subspace. The Schwarz inequality shows that

$$|(h,f) - (h,g)| = |(h, f-g)| \leq \|h\| \|f-g\|$$

and so for any given h, the mapping F defined by

$$F(f) = (h, f)$$

is continuous (Appendix p. 228). Now $F^{-1}(0)$ is the set of all f in L^2 for which $F(f) = 0$. In other words $F^{-1}(0) = h^\perp$. As F is continuous, and $\{0\}$ is a closed subset of $\mathbf{R}$, it follows that $h^\perp$ is a closed subset of L^2 (Appendix, Proposition 3). The linear subspace $S^\perp$ is the intersection of all the closed sets $s^\perp$ for s in S and so $S^\perp$ *is a closed linear subspace of* L^2.

Theorem 2 allows us to prove a result of fundamental importance for L^2 (and Hilbert Space generally, see §7.7).

Theorem 3 (*The Projection Theorem*).

Let M be a closed linear subspace of L^2. Then any element f of L^2 may be expressed uniquely in the form

$$f = g + h, \tag{3}$$

where $g \in M$, $h \in M^\perp$.

It is customary to call g, h, respectively, the *(orthogonal) projections* of f on M, $M^\perp$. Once again the uniqueness is strictly speaking in $\mathscr{L}^2$, i.e. to within null functions.

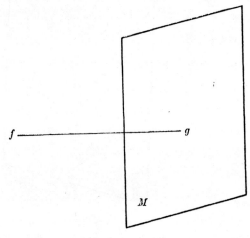

Fig. 64

Proof. Let g be the nearest point of M to f and define $h = f - g$, where $h \in M^\perp$ by Theorem 2. This gives the required expression for f.
Now
$$M \cap M^\perp = \{0\}$$
because
$$(f,f) = 0 \Rightarrow f = 0 \text{ (a.e.)}.$$
The uniqueness follows, for if f also equals $g_1 + h_1$, where $g_1 \in M$, $h_1 \in M^\perp$, then
$$g - g_1 = h_1 - h \in M \cap M^\perp.$$

In view of this result, $M^\perp$ is often called the *orthogonal complement* of M. By the uniqueness of the expression (3) it follows that M is the orthogonal complement of $M^\perp$; in other words
$$M^{\perp\perp} = M.$$

Exercises

1. Write out the proofs of Propositions 1, 2.

2. If $h = (1-t)f + tg$ $(0 \leq t \leq 1)$ verify that
$$d(f,h) : d(h,g) = t : (1-t).$$
How does this extend if $t < 0$ or $t > 1$?

3. If the parallelogram law is satisfied in L^p show that $p = 2$.

4. Let
$$f(x) = (1+x^{1/p})^p + (1-x^{1/p})^p$$
for $0 < x < 1$.

(i) If $p \geq 2$ show that $f''(x) \leq 0$ for $0 < x < 1$ and deduce that
$$f(x) \leq f(c) + (x-c)f'(c)$$
for $0 < x, c < 1$. Now let $c \uparrow 1$ and deduce that
$$f(x) \leq 2^{p-1}(x+1)$$
for $0 < x < 1$.

Let p, q satisfy the relation $(1/p) + (1/q) = 1$. Substitute $c = x^q$ and deduce that
$$f(x) \leq 2(1+x^{q/p})^{p/q}$$
for $0 < x < 1$.

(ii) If $1 < p \leq 2$ show that these inequalities hold in the reverse sense.

5. (i) Use the inequalities of Ex. 4 to prove that, if $p \geq 2$, then
$$2\{|a|^p + |b|^p\} \leq |a+b|^p + |a-b|^p \leq 2^{p-1}\{|a|^p + |b|^p\},$$
$$|a+b|^p + |a-b|^p \leq 2\{|a|^q + |b|^q\}^{p/q},$$
$$2\{|a|^p + |b|^p\}^{q/p} \leq |a+b|^q + |a-b|^q$$

for all real numbers a, b $(1/p + 1/q = 1)$. Deduce *Clarkson's Inequalities* [5] for L^p ($p \geq 2$):
$$2\{\|f\|^p + \|g\|^p\} \leq \|f+g\|^p + \|f-g\|^p \leq 2^{p-1}\{\|f\|^p + \|g\|^p\},$$
$$\|f+g\|^p + \|f-g\|^p \leq 2\{\|f\|^q + \|g\|^q\}^{p/q},$$
$$2\{\|f\|^p + \|g\|^p\}^{q/p} \leq \|f+g\|^q + \|f-g\|^q \quad (f, g \in L^p).$$

(ii) Show that these inequalities hold in the reverse sense if $1 < p \leq 2$. (Recall from Ex. 7.2.5 that Minkowski's Inequality is reversed for L^s when $0 < s \leq 1$.)

6. Let S be a closed convex set in L^p ($p > 1$). Use Clarkson's Inequalities (Ex. 5) to prove that S contains a unique element of smallest norm. Does this result hold for L^1?

7. In the notation of the Projection Theorem write $Pf = g$, $Qf = h$. Show that P, Q are continuous linear mappings of L^2 onto M, $M^\perp$ respectively, and that
$$P^2 = P, \quad Q^2 = Q, \quad PQ = QP = 0.$$

7.4 Bounded linear functionals on L^2

In the previous section we considered the function $F: L^2 \to \mathbf{R}$ defined, for a particular h in L^2, by
$$F(f) = (h, f).$$

From the fundamental rules SP 1, SP 2:

$$F(af+bg) = aF(f)+bF(g)$$

for any f, g in L^2 and any a, b in **R**. Thus F is a *linear* function: historically such a function F on L^2 with values in **R** was called a *linear functional* as it was thought expedient to distinguish F from the other functions $f, g, \ldots$ on which F operates. This name has remained with us, although it is now used for real (or complex) valued linear functions on abstract linear spaces, whose elements may, or may not, be functions.

The fact that F is continuous follows from Schwarz' Inequality because
$$|F(f)| \leq \|h\| \|f\|$$
for all f, and so
$$|F(f-g)| \leq \|h\| \|f-g\|.$$

A linear functional F on L^2 for which there is a positive real number K satisfying
$$|F(f)| \leq K\|f\| \tag{1}$$
for all f, is called a *bounded linear functional*.

Proposition 1. *A linear functional on L^2 is bounded if and only if it is continuous.*

Proof. (i) If F satisfies inequality (1) then
$$|F(f-g)| \leq K\|f-g\|$$
which immediately gives continuity.

(ii) Suppose that F is continuous – in particular at 0. Given $\epsilon > 0$; there exists $\delta > 0$ such that
$$|F(g)| < \epsilon \quad \text{whenever} \quad \|g\| < \delta.$$

For any f with $\|f\| \neq 0$ let
$$g = \frac{\delta}{2\|f\|} f.$$

Then
$$\|g\| = \tfrac{1}{2}\delta$$

and so
$$|F(g)| < \epsilon.$$

By the linearity of F this means that
$$|F(f)| < (2\epsilon/\delta)\|f\|.$$

Thus (1) is satisfied with $K = 2\epsilon/\delta$ (even if $\|f\| = 0$).

As a simple consequence of the geometric reasoning of §7.3 we now deduce a very famous result [20].

BOUNDED LINEAR FUNCTIONALS

Theorem 1 (Riesz). *If F is a bounded linear functional on L^2, then there exists an element h of L^2 such that*

$$F(f) = (h,f)$$

for all f in L^2, and h is unique to within a null function.

We may take our lead from the familiar picture in $\mathbf{R}^3$ where the linear equation

$$a_1 x_1 + a_2 x_2 + a_3 x_3 = 0,$$

i.e. $\qquad a.x = 0$

defines a plane, provided not all the a's are zero. The coefficients a_1, a_2, a_3 are determined by this plane to within a constant multiple, as the direction of a is perpendicular to the plane (Fig. 65).

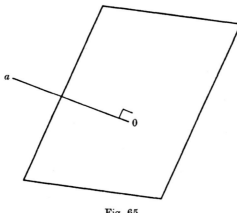

Fig. 65

Proof. If F is the zero functional we may obviously take $h = 0$. Suppose therefore that F is not the zero functional and so the linear subspace

$$M = \{f \in L^2 : F(f) = 0\}$$

is not the whole of L^2. But F is continuous, thus M is closed, and the Projection Theorem yields a non-zero orthogonal complement $M^\perp$. In fact $M^\perp$ is one-dimensional, for if $h, h' \in M^\perp$ and $F(h) \neq 0$, we may choose a real number t so that

$$F(h' - th) = 0,$$

whence $\qquad h' - th \in M \cap M^\perp = \{0\}.$

If we multiply h by a suitable non-zero number we may also satisfy the condition

$$F(h) = (h, h)$$

(which is obviously necessary). This non-zero element h of $M^\perp$ now solves our problem, for, by the Projection Theorem, an arbitrary element f of L^2 may be expressed as
$$f = m + th,$$
where $m \in M$, $t \in \mathbf{R}$, and so
$$F(f) = tF(h) = t(h,h) = (h,f).$$
The uniqueness of h is clear: if
$$(h,f) = (h',f)$$
for all f, then $\qquad h - h' \in M \cap M^\perp = \{0\}.$

The bounded linear functionals on L^2 themselves form a linear space over $\mathbf{R}$ if we define $aF + bG$ by the obvious rule
$$(aF + bG)(f) = aF(f) + bG(f).$$
Moreover, a bounded linear functional F on L^2 may be given a *norm* in the following way: there is a positive real number K such that
$$|F(f)| \leqslant K\|f\|$$
for all f in L^2, i.e. K is an upper bound for the set
$$\{|F(f)|/\|f\| : f \in L^2, \|f\| \neq 0\}.$$
Let $\|F\|$ be the smallest such number K, i.e.
$$\|F\| = \sup\{|F(f)|/\|f\| : f \in L^2, \|f\| \neq 0\}.$$
In view of the linearity of F this is easily seen to be equivalent to
$$\|F\| = \sup\{F(f) : f \in L^2, \|f\| \leqslant 1\}.$$
From this definition it follows that
$$\|F\| \geqslant 0,$$
$$\|F\| = 0 \quad \text{if and only if} \quad F = 0,$$
$$\|aF\| = |a|\,\|F\|,$$
$$\|F + G\| \leqslant \|F\| + \|G\|,$$
the last mentioned depending only on the triangle inequality
$$|F(f) + G(f)| \leqslant |F(f)| + |G(f)|$$
for real numbers. The normed linear space of bounded linear functionals on L^2 is called the *dual space* of L^2. In view of Riesz's Theorem 1 we can now show that this dual space is virtually the same as L^2, and we may loosely say that L^2 is *self-dual*.

Theorem 2. *For any h in L^2 let F_h denote the bounded linear functional defined by*
$$F_h(f) = (h,f) \quad (f \in L^2).$$
Then the mapping
$$\theta : h \to F_h$$
is an isomorphism of the normed linear space L^2 onto its dual space.

By 'isomorphism' here we mean a one-one linear mapping which preserves the norm, in the sense that
$$\|F_h\| = \|h\|$$
for all h in L^2.

Proof. The linearity of θ is immediate from the definition of F_h and the linear properties of the scalar product; the fact that θ is one-one and onto is the substance of Riesz's Theorem.

It only remains to check that $\|F_h\| = \|h\|$. From Schwarz' Inequality
$$|F_h(f)| \leq \|h\|\|f\|$$
and so
$$\|F_h\| \leq \|h\|.$$
But also
$$|F_h(h)| = \|h\|^2,$$
from which
$$\|F_h\| \geq \|h\|.$$
This completes the proof.

It is worth noting that the isomorphism θ of Theorem 2 is 'canonical' in the sense that it does not depend on any frame of reference, or basis, in L^2 – in fact we have given no definition, let alone the construction, of such a basis. The definition we gave of the dual space of L^2 carries over virtually unchanged to L^p for $p > 1$. There is a deeper theorem to the effect that, if $(1/p) + (1/q) = 1$, the normed linear spaces L^p, L^q are duals of each other (to within a canonical isomorphism as in Theorem 2). We shall prove this result in the sequel as an application of the Radon–Nikodym Theorem. For a geometric proof on the lines of the one given above for L^2 see Exx. 5–10.

Riesz's Theorem states that for any bounded linear functional F on L^2 there is a function h in L^2 such that
$$F(f) = \int fh$$
for all f in L^2. This is strikingly reminiscent of Theorem 6.4.1, on which we based our discussion of the transformation of integrals on $\mathbf{R}^k$. Recall that T is a continuous one-one mapping of an open subset U of $\mathbf{R}^k$ onto a subset V of $\mathbf{R}^k$ and that T has a *density function* h satisfying
$$m(TI) = \int_I h$$

for all intervals I in $\mathbf{R}^k$ with $\bar{I} \subset U$. Let L denote the linear space of functions $f\colon U \to \mathbf{R}$ for which $fh \in L^1(U)$ and let $g\colon V \to \mathbf{R}$ be defined by
$$g(Tx) = f(x)$$
for all x in U. Theorem 6.4.1 then asserts that $g \in L^1(V)$ for all f in L and
$$\int_V g = \int_U fh.$$
If we care to define the *positive linear functional* F on L by the equation
$$F(f) = \int_V g,$$
then the conclusion reads
$$F(f) = \int_U fh$$
for all f in L. In the sequel we shall prove the Radon–Nikodym Theorem which gives a necessary and sufficient condition for the existence of such density functions in a very general setting; the proof is based on Riesz's Theorem for L^2.

Exercises

1. Let F be the bounded linear functional on $\mathbf{R}^3$ defined by the equation
$$F(x) = a_1 x_1 + a_2 x_2 + a_3 x_3.$$
Show that $\|F\| = (a_1^2 + a_2^2 + a_3^2)^{\frac{1}{2}}$. In the case where $\|F\| = 1$ show that (a_1, a_2, a_3) is the point of the plane
$$a_1 x_1 + a_2 x_2 + a_3 x_3 = 1$$
nearest to the origin.

2. Let F be a bounded linear functional on L^2 for which $\|F\| = 1$. Show that there is a unique point h of L^2 on the unit sphere $\{f \in L^2 \colon \|f\| = 1\}$ such that $F(h) = 1$. Show that h is also the unique point of the 'hyperplane' $\{f \in L^2 \colon F(f) = 1\}$ nearest to the origin.

3. Let F and h be as in Ex. 2. If $F(m) = 0$, show that $\|h + tm\|^2$ attains a minimum at $t = 0$; by differentiating and substituting $t = 0$ show that $(h, m) = 0$. Hence show that $F(f) = (h, f)$ for all f in L^2. (Cf. the proof of Theorem 1.)

4. (An alternative to Ex. 3.) Let F and h be as in Ex. 2 and define
$$\phi_f(t) = \|h + tf\|$$
for each f in L^2. From the inequality
$$F(h + tf) \leq \|h + tf\|$$

deduce that, if $\phi_f'(0)$ exists, then
$$\phi_f'(0) = F(f).$$
Hence prove that $\quad F(f) = (h,f)$
for all f in L^2.

The Duality of the spaces L^p, L^q

In the following exercises $p, q > 1$ and $(1/p) + (1/q) = 1$.

5. To any element h of L^p there corresponds an element
$$g = |h|^{p-1} \operatorname{sgn} h.$$
Show that $g \in L^q$ and that $\quad h = |g|^{q-1} \operatorname{sgn} g.$
Show also that the following 'symmetric' relations hold:
$$hg = |h|^p = |g|^q$$
$$\int hg = \|h\| \|g\|.$$
(Cf. Hölder's Inequality.)

6. To any g in L^q there is a linear functional F on L^p defined by
$$F(f) = \int gf.$$
By considering the pair (h, g) of elements of Ex. 5 show that
$$\|F\| = \|g\|.$$
Show also that h is the element of smallest norm in the 'hyperplane'
$$\{f \in L^p : F(f) = \|F\|^q\}.$$

7. (Cf. Ex. 2 and Ex. 7.3.6.) Let F be a bounded linear functional on L^p for which $\|F\| = 1$. Show that there is a unique point h of L^p on the unit sphere $\{f \in L^p : \|f\| = 1\}$ such that $F(h) = 1$. Show that h is the unique point of the 'hyperplane' $\{f \in L^p : F(f) = 1\}$ nearest to the origin.

8. (Cf. Ex. 3.) Let F and h be as in Ex. 7.
(i) If $F(m) = 0$ show that $\int |h + tm|^p$ attains a minimum at $t = 0$.
(ii) For any real numbers a, b let
$$\psi(t) = |a + tb|^p = (a^2 + 2tab + t^2b^2)^{\frac{1}{2}p}.$$
Show that $\quad \psi'(t) = p|a + tb|^{p-2}(a + tb) b.$

(If $a + tb = 0$ the right hand side is interpreted as zero even if $p < 2$: here we definitely need $p > 1$.)

(iii) Justify differentiating $\int |h+tm|^p$ under the integral sign (Exx. 5.2.12, 13) and substituting $t = 0$ to show that

$$\int gm = 0,$$

where g is related to h as in Ex. 5.

(iv) Show that
$$F(f) = \int gf$$
for all f in L^p.

9. (A variation on Ex. 8. Cf. Ex. 4.) Let F and h be as in Ex. 7 and define

$$\phi_f(t) = \|h+tf\|$$

for each f in L^p. From the inequality

$$F(h+tf) \leq \|h+tf\|$$

deduce that, if $\phi_f'(0)$ exists, then

$$\phi_f'(0) = F(f).$$

Justify the differentiation under the integral sign as in Ex. 8 (iii) and hence prove that
$$F(f) = \int gf$$
for all f in L^p.

10. For any g in L^q define the bounded linear functional F_g on L^p by the rule
$$F_g(f) = \int gf.$$

Use Exx. 6, 8 to show that $\quad \theta: g \to F_g$

is an isomorphism of the normed linear space L^q onto the dual space of L^p.

7.5 Orthonormal sets in L^2

In many branches of mathematics it is important to 'expand' a function f in terms of certain given functions whose behaviour is well known. Perhaps the most famous example of all is the Fourier expansion of $f(x)$ as a trigonometric series

$$\tfrac{1}{2}a_0 + a_1 \cos x + b_1 \sin x + a_2 \cos 2x + b_2 \sin 2x + \dots.$$

(The reason for the $\tfrac{1}{2}$ in front of a_0 will appear shortly.) Quite generally, if a sequence $\{f_n\}$ of functions is given, we shall be interested in finding the series
$$c_1 f_1 + c_2 f_2 + \dots$$
which most nearly approximates f in the sense of a norm, or distance.

To begin with we consider finite expansions. Let $f_1, f_2, \ldots, f_m$ be functions in L^2 and denote by M the set of all linear combinations

$$a_1 f_1 + \ldots + a_m f_m$$

with $a_1, \ldots, a_m$ in $\mathbf{R}$. It is clear that M is a linear space; we say that M is *spanned* by $f_1, \ldots, f_m$. In terms of coordinates, the 'dropping of perpendiculars' onto M is greatly simplified if we further assume that $f_1, \ldots, f_m$ are mutually orthogonal, i.e. if we assume that

$$(f_i, f_j) = 0 \quad \text{for} \quad i \neq j. \tag{1}$$

Theorem 1. *Let M be the linear space spanned by the non-zero orthogonal functions $f_1, \ldots, f_m$ in L^2. To any point f of L^2 there is a unique nearest point g of M, viz. the orthogonal projection of f on M. In other words there is a unique best approximation*

$$g = c_1 f_1 + \ldots + c_m f_m$$

for f as a linear combination of $f_1, \ldots, f_m$, and the coefficients $c_1, \ldots, c_m$ are uniquely determined by the equations

$$c_i(f_i, f_i) = (f, f_i) \quad (i = 1, \ldots, m). \tag{2}$$

These coefficients c_i are often called the *Fourier coefficients* of f with respect to $\{f_1, \ldots, f_m\}$.

Proof. The condition $f - g \in M^\perp$ is equivalent to

$$(f - g, f_i) = 0 \quad (i = 1, \ldots, m)$$

and this in turn is equivalent to (2) in view of the orthogonality relations (1). In particular, equations (2) show that g is the only point of M satisfying the condition $f - g \in M^\perp$. Moreover, g is the unique nearest point of M to f, for if e is a point of M distinct from g, we may apply Pythagoras' Theorem to the right angled triangle efg whose hypotenuse ef is longer than the side fg (Fig. 66).

For example, in $L^2(-\pi, \pi)$ the trigonometric functions satisfy the famous orthogonality relations

$$\left. \begin{aligned} \int_{-\pi}^{\pi} \cos mx \cos nx \, dx &= 0 \quad (m \neq n), \\ \int_{-\pi}^{\pi} \sin mx \sin nx \, dx &= 0 \quad (m \neq n), \\ \int_{-\pi}^{\pi} \cos mx \sin nx \, dx &= 0 \end{aligned} \right\} \tag{3}$$

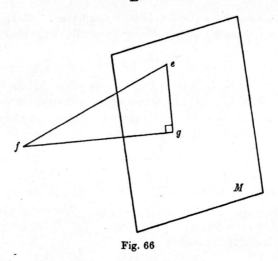

Fig. 66

and
$$\int_{-\pi}^{\pi} \cos^2 nx\, dx = 2\pi \quad \text{if} \quad n = 0,$$
$$= \pi \quad \text{if} \quad n \geqslant 1;$$
$$\int_{-\pi}^{\pi} \sin^2 nx\, dx = \pi \quad \text{if} \quad n \geqslant 1. \qquad (4)$$

Thus the finite expansion

$$\tfrac{1}{2}a_0 + a_1 \cos x + b_1 \sin x + \ldots + a_N \cos Nx + b_N \sin Nx,$$

or *trigonometric polynomial of degree* N, as it is often called in the classical literature, is a 'best possible' expansion of $f(x)$ in this form, provided the coefficients are given by

$$a_n = \frac{1}{\pi} \int_{-\pi}^{\pi} f(x) \cos nx\, dx \quad (n \geqslant 0),$$
$$b_n = \frac{1}{\pi} \int_{-\pi}^{\pi} f(x) \sin nx\, dx \quad (n \geqslant 1). \qquad (5)$$

The reason for the $\tfrac{1}{2}$ in front of a_0 is now clear; it allows the above unified formula for the a's to include the case $n = 0$. The classical pointwise notation for functions should not be allowed to obscure the fact that the above trigonometric polynomial is *best possible in the sense of the norm*; for certain points x it may be a very bad approximation indeed!

If $f_1, \ldots, f_m$ are arbitrary elements of L^2 then we may orthogonalise these functions in a natural way: each simple step of the construction involves the orthogonal projection of f_{r+1} on the linear space spanned by $f_1, \ldots, f_r$.

Theorem 2 (Gram–Schmidt). *Let $f_1, \ldots, f_m \in L^2$. Then there exist orthogonal functions $h_1, \ldots, h_m$ in L^2 such that, for $r = 1, \ldots, m$, $\{f_1, \ldots, f_r\}$ and $\{h_1, \ldots, h_r\}$ span the same linear subspace of L^2.*

Proof. Suppose that orthogonal functions $h_1, \ldots, h_r$ have been found which span the same linear space as $f_1, \ldots, f_r$. If $r < m$ we may define
$$h_{r+1} = f_{r+1} - a_1 h_1 - \ldots - a_r h_r$$
uniquely by insisting that h_{r+1} be orthogonal to h_i for $i = 1, \ldots, r$. If $h_i \neq 0$ this condition requires
$$(f_{r+1}, h_i) = a_i(h_i, h_i)$$
which determines a_i uniquely; and if $h_i = 0$ it does not matter what real value we give to a_i (for definiteness let $a_i = 0$ in this case). The orthogonal functions $h_1, \ldots, h_{r+1}$ now span the same linear space as $f_1, \ldots, f_{r+1}$. The proof follows at once by induction if we begin by setting $h_1 = f_1$.

This proof differs slightly from the standard one in that we have allowed zero functions to occur among the h's. In fact, all the f's may be zero, in which case all the h's would also be zero. Apart from this trivial case we may omit any zero function from among the h's and 'normalise' the remaining h_i by dividing by $\|h_i\|$. In this way we obtain elements $e_1, \ldots, e_n$ of L^2 which are *orthonormal*, i.e. satisfy the conditions
$$\begin{aligned}(e_i, e_j) &= 0 \quad \text{if} \quad i \neq j, \\ &= 1 \quad \text{if} \quad i = j.\end{aligned} \tag{6}$$

For simplicity of expression we also say that the set $\{e_1, \ldots, e_n\}$ is orthonormal. As a convention we agree that the zero subspace $\{0\}$ is spanned by the empty set $\varnothing$, and that $\varnothing$ is orthonormal in the sense that it has no elements which violate the conditions (6). A *finite dimensional* subspace M of L^2 is one which is spanned by a finite set of elements of L^2. By the above convention this definition includes the zero subspace. With this terminology we now have:

Corollary. *Any finite dimensional subspace M of L^2 may be spanned by an orthonormal set.*

In view of this result we may apply Theorem 1 to any finite dimensional subspace M and express the orthogonal projection g of f on M as
$$g = c_1 e_1 + \ldots + c_n e_n,$$

where the coefficients are given by the simplified equations

$$c_i = (f, e_i) \quad (i = 1, \ldots, n) \tag{7}$$

in place of (2).

In particular, taking $f = 0$, we see that the only linear relation

$$a_1 e_1 + \ldots + a_n e_n = 0$$

is the trivial one with $a_1 = \ldots = a_n = 0$; we say that $e_1, \ldots, e_n$ are *linearly independent*. In general, we say that $f_1, \ldots, f_m$ are *linearly dependent* if there is a linear relation

$$a_1 f_1 + \ldots + a_m f_m = 0$$

with not all $a_i = 0$, and *linearly independent* otherwise. A linearly independent spanning set for M is called a *basis* of M. If $\{f_1, \ldots, f_m\}$ is such a basis, then any f in M is uniquely expressible as

$$f = c_1 f_1 + \ldots + c_m f_m,$$

where we think of $c_1, \ldots, c_m$ as coordinates of f with respect to the basis (Ex. 1). The advantage of an orthonormal basis $\{e_1, \ldots, e_n\}$ is that the coordinates c_i are so easily obtained by orthogonal projection using equation (7).

If $\{f_1, \ldots, f_m\}$ is a basis of M then the orthogonal elements $h_1, \ldots, h_m$ given by the Gram–Schmidt process are all non-zero: first of all $h_1 = f_1 \neq 0$; and if $h_{r+1} = 0$ we should have

$$f_{r+1} = a_1 h_1 + \ldots + a_r h_r$$

which is a linear combination of $f_1, \ldots, f_r$ by construction. Thus the orthonormal basis $\{e_1, \ldots, e_n\}$ in this case has $n = m$. This suggests that we define the *dimension* of a finite dimensional space M as the number of elements in any basis. To justify this definition we need a reassurance that any two bases have the same number of elements.

Lemma 1. *If $g_1, \ldots, g_n$ are linearly independent elements in the linear space M spanned by $f_1, \ldots, f_m$, then $n \leq m$.*

Proof. As $g_1 \in M$,

$$g_1 = a_1 f_1 + \ldots + a_m f_m,$$

where not all the a's are zero (otherwise $g_1 = 0$ would contradict the linear independence of $g_1, \ldots, g_n$). By rearranging the order of the f's, if necessary, we may assume that $a_1 \neq 0$ and so f_1 may be expressed as a linear combination of $g_1, f_2, \ldots, f_m$. This means that the linear space spanned by $g_1, f_2, \ldots, f_m$ contains $f_1, f_2, \ldots, f_m$ and so must be M itself.

Now $g_2 \in M$ and so
$$g_2 = b_1 g_1 + c_2 f_2 + \ldots + c_m f_m,$$
where not all the c's are zero (otherwise $g_2 = b_1 g_1$ would contradict the linear independence of $g_1, \ldots, g_n$). By further rearranging the order of $f_2, \ldots, f_m$, if necessary, we may assume that $c_2 \neq 0$ and so f_2 may be expressed as a linear combination of $g_1, g_2, f_3, \ldots, f_m$. The linear space spanned by $g_1, g_2, f_3, \ldots, f_m$ contains $g_1, f_2, f_3, \ldots, f_m$ and so must be M again. This process may be continued as long as there are any g's remaining, and so $n \leq m$.

As a significant bonus we also note that M is spanned by $g_1, \ldots, g_n$ together with a suitable selection of $m - n$ of the f's. In this form the result is often referred to as the *Exchange Lemma* (see Exx. 3, 4).

To prove the consistency of our definition of (finite) dimension we observe that if $\{f_1, \ldots, f_m\}$, $\{g_1, \ldots, g_n\}$ are both bases of M, then Lemma 1 applies two ways to give $n \leq m$ and $m \leq n$.

So far in this section we have not used the completeness of L^2, and in particular we have not deduced Theorem 1 from the deeper geometric theorems of §7.3. In view of Theorem 2, Corollary and Theorem 1, any finite dimensional subspace M of L^2 has the property that, if $f \notin M$, then there is a nearest point g of M to f which satisfies $\|f - g\| > 0$. This shows that the complement of M in L^2 is open. Thus

any finite dimensional linear subspace of L^2 is closed.

As an alternative to the above treatment, one may prove this last mentioned result at the beginning, and then deduce Theorem 1 from it (see Ex. 6).

As we move on to the study of infinite expansions the completeness of L^2 plays a dominant role. Let $f_1, f_2, \ldots$ belong to L^2. The finite expansions
$$c_1 f_1 + \ldots + c_m f_m$$
(for arbitrary m) form a linear space M; we say that M is *spanned* by $f_1, f_2, \ldots$. As we are interested in convergent series
$$c_1 f_1 + c_2 f_2 + \ldots$$
we shall have to consider the *closure* $\bar{M}$ of M, i.e. the smallest closed subset of L^2 containing M (Appendix).

Lemma 2. *The closure $\bar{M}$ of M is a linear subspace of L^2.*

Proof. Recall from the Appendix that $\bar{M}$ consists of the adherent points of M (Appendix, Ex. 2). Let $f, g \in \bar{M}$ and $a, b \in \mathbf{R}$. To any $\epsilon > 0$, there exist $f', g' \in M$ such that $\|f-f'\|$ and $\|g-g'\|$ are both less than $\epsilon/(|a|+|b|+1)$. Thus

$$\|(af+bg)-(af'+bg')\| \leq |a|\|f-f'\|+|b|\|g-g'\| < \epsilon.$$

As $af'+bg' \in M$ and ϵ is arbitrarily small, this means that $af+bg \in \bar{M}$.

Lemma 3. $M^\perp = \bar{M}^\perp$.

Proof. As $M \subset \bar{M}$ the inclusion $\bar{M}^\perp \subset M^\perp$ is obvious. Let $f \in M^\perp$, $g \in \bar{M}$. To any $\epsilon > 0$ there exists g' in M such that $\|g-g'\| < \epsilon$. By Schwarz' Inequality

$$|(f,(g-g'))| \leq \|f\|\epsilon.$$

As $(f,g') = 0$ this means that

$$|(f,g)| \leq \|f\|\epsilon$$

for arbitrarily small $\epsilon > 0$. Thus $(f,g) = 0$. This gives the reverse inclusion $M^\perp \subset \bar{M}^\perp$ and completes the proof.

We shall say that $f_1, f_2, \ldots$ are *linearly independent* if $f_1, \ldots, f_m$ are linearly independent of every $m \geq 1$. For example, if $\{I_n\}$ is a sequence of disjoint intervals in $\mathbf{R}$ with $m(I_n) > 0$, then the corresponding characteristic functions

$$\chi_{I_1}, \chi_{I_2}, \ldots$$

are linearly independent in L^2. (Remember that null functions are regarded as zero functions, and so we must avoid $m(I_n) = 0$.) If we apply the Gram–Schmidt process to $f_1, f_2, \ldots$ we obtain orthogonal functions $h_1, h_2, \ldots$ none of which is zero (by the argument given just before Lemma 1). The h's normalise to give an orthonormal set $\{e_1, e_2, \ldots\}$ in L^2.

If $f \in L^2$, the *Fourier series* of f with respect to the orthonormal set $\{e_1, e_2, \ldots\}$ is

$$c_1 e_1 + c_2 e_2 + \ldots,$$

where
$$c_n = (f, e_n) \quad (n = 1, 2, \ldots).$$

We now have the deeply satisfying result that *any* f in L^2 has a convergent Fourier series with respect to *any* orthonormal set. Admittedly this Fourier series need not converge to f, but it does converge to the nearest 'available' point.

Theorem 3. *Let M be the linear space spanned by an orthonormal set $\{e_1, e_2, \ldots\}$ in L^2 and $\bar{M}$ the closure of M. If $f \in L^2$, then the Fourier series of f converges to the orthogonal projection g of f on $\bar{M}$; in particular, the Fourier series of f converges to f if and only if $f \in \bar{M}$.*

It is most important to stress once again that convergence here is convergence in mean, and that this is not the same as pointwise convergence.

Proof. As usual $$c_n = (f, e_n)$$
for $n = 1, 2, \ldots$ and the n-th partial sum of the Fourier series of f is
$$g_n = c_1 e_1 + \ldots + c_n e_n.$$
By Theorem 1, g_n is the orthogonal projection of f on the finite dimensional subspace M_n spanned by $e_1, \ldots, e_n$ (Fig. 67). By Pythagoras' Theorem
$$\|g_n\|^2 = c_1^2 + \ldots + c_n^2 \leq \|f\|^2.$$

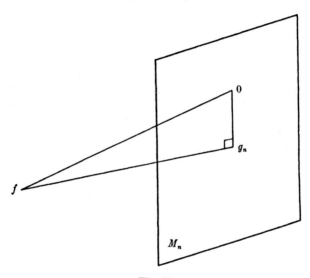

Fig. 67

Thus the series Σc_n^2 is convergent and the sum
$$\sum_{n=1}^{\infty} c_n^2 \leq \|f\|^2.$$
Given $\epsilon > 0$; we can find N so that
$$\sum_{N+1}^{\infty} c_n^2 < \epsilon^2,$$
whence
$$\|g_m - g_n\|^2 = c_{n+1}^2 + \ldots + c_m^2 < \epsilon^2 \quad \text{for} \quad m > n \geq N.$$
Therefore $\{g_n\}$ is a Cauchy sequence and the completeness of L^2 guarantees the convergence of $\{g_n\}$ to an element g of L^2. As $g_n \in M$

for all n, $g \in \bar{M}$. Also $f - g_n \in M_n^\perp \subset M_N^\perp$ for $n \geq N$. As $M_N^\perp$ is closed, $f - g \in M_N^\perp$ (for $N \geq 1$) and so $f - g \in M^\perp = \bar{M}^\perp$ by Lemma 3. Thus g is the orthogonal projection of f on $\bar{M}$.

In view of the last clause of Theorem 3 it is of considerable interest to know whether or not $\bar{M}$ is the whole of L^2. Let M be the linear subspace spanned by a set $\{f_1, f_2, \ldots\}$ in L^2; if M is *dense* in L^2, i.e. if $\bar{M} = L^2$, we shall say that $\{f_1, f_2, \ldots\}$ is *complete* in L^2. (The word 'complete' is unfortunate here but is deeply entrenched in the literature.)

Theorem 4. *There exist complete orthonormal sets in L^2.*

Proof. The proof proceeds in several steps.

(i) *For any f in L^2 and any $\epsilon > 0$, we may find a step function ϕ such that $\|f - \phi\| < \epsilon$.*

To prove this consider first the case where f is positive and let
$$f_n = \min\{f, nf^2, n\}$$
for $n \geq 1$. Then
$$f_n \in L^2 \quad \text{as} \quad f_n^2 \leq f^2 \quad \text{and} \quad (f - f_n)^2 \downarrow 0.$$

By the Monotone Convergence Theorem
$$\|f - f_n\| \to 0.$$

Choose n so large that $\quad \|f - f_n\| < \epsilon/2$.

Now $f_n \in L^1$ as $f_n \leq nf^2$ and so, by Ex. 3.2.5, we may find a step function ϕ such that
$$\int |f_n - \phi| < \epsilon^2/4n.$$

As $0 \leq f_n \leq n$ we may truncate ϕ, if necessary, to ensure that
$$0 \leq \phi \leq n.$$
This gives
$$\|f_n - \phi\|^2 = \int |f_n - \phi| |f_n - \phi| \leq n \int |f_n - \phi| < \epsilon^2/4$$
and so $\quad \|f_n - \phi\| < \epsilon/2$.

By the triangle inequality
$$\|f - \phi\| \leq \|f - f_n\| + \|f_n - \phi\| < \epsilon.$$

The result extends to any f in L^2 by applying the triangle inequality to $f = f^+ - f^-$.

(ii) *For any step function ϕ and any $\epsilon > 0$, we may find a rational step function ψ such that $\|\phi - \psi\| < \epsilon$.*

By a rational step function we mean one whose discontinuities, if any, occur at rational points. This is clear if we express ϕ in terms of disjoint intervals I_r and approximate each I_r (at least when $m(I_r) > 0$) by means of a rational interval $J_r \subset I_r$. (At this stage we could also insist that ψ should take only rational values, but when we 'normalise' in (iii) we may introduce irrational values as square roots.)

(iii) *There exists a complete orthonormal set $\{\psi_1, \psi_2, \ldots\}$ of rational step functions.*

The rational intervals may be arranged as the terms of a sequence $\{J_n\}$ (cf. §2.1). If we apply the Gram–Schmidt process to the sequence

$$\chi_{J_1}, \chi_{J_2}, \ldots$$

of characteristic functions of these intervals, suppress any null functions that arise, and normalise the remaining functions, we obtain an orthonormal sequence

$$\psi_1, \psi_2, \ldots$$

of rational step functions. The linear space M spanned by $\psi_1, \psi_2, \ldots$ contains all χ_{J_n} and hence contains all rational step functions.

To any f in L^2 and any $\epsilon > 0$, we may find a step function ϕ as in (i), then a rational step function ψ as in (ii) so that

$$\|f - \psi\| \leq \|f - \phi\| + \|\phi - \psi\| < 2\epsilon.$$

But ψ belongs to the linear space M, and so M is dense in L^2, as required.

There are two observations about the 'size' of orthonormal sets.

Proposition 1. (i) *Any complete orthonormal set in L^2 is infinite.*

(ii) *Any orthonormal set in L^2 is countable.*

Proof. (i) If $\{e_1, \ldots, e_n\}$ were complete, the (closed) linear subspace spanned by $e_1, \ldots, e_n$ would coincide with L^2. But L^2 is infinite dimensional (i.e. not finite dimensional): consider, for example, the characteristic functions of a sequence of disjoint intervals $\{I_n\}$ with

$$m(I_n) > 0,$$

and apply Lemma 1.

(ii) Let S be an orthonormal set in L^2. By Pythagoras' Theorem the distance between any two points of S is $\sqrt{2}$. For each f in S we may find a *rational step function θ with rational values* such that

$$\|f - \theta\| < \tfrac{1}{2}\sqrt{2}. \tag{8}$$

(Adapt part (ii) of the proof of Theorem 4 by allowing ψ to take only rational values.) Now the collection of all such θ may be written as the terms of a sequence $\{\theta_n\}$ (cf. §2.1, again). For definiteness, choose θ_{n_f} to be the first of these to satisfy (8). If $f, g \in S$ and $n_f = n_g$ then

$$\|f - g\| \leq \|f - \theta_{n_f}\| + \|\theta_{n_g} - g\| < \sqrt{2}$$

and so $f = g$. Thus the mapping

$$f \to n_f$$

is a one-one mapping of S into $\{1, 2, 3, \ldots\}$ and this gives the result.

In view of Theorem 4 there is now a meaningful special case of Theorem 3.

Theorem 5. *Let $\{e_1, e_2, \ldots\}$ be a complete orthonormal set in L^2. For any f in L^2 the Fourier series of f with respect to $\{e_1, e_2, \ldots\}$ converges in mean to f.*

This means that, once a complete orthonormal set $\{e_1, e_2, \ldots\}$ is chosen, any f in L^2 may be described in terms of the sequence $\{c_n\}$ of Fourier coefficients given by

$$c_n = (f, e_n) \quad (n = 1, 2, \ldots).$$

These coefficients are not quite arbitrary. Recall from the proof of Theorem 3 that

$$\Sigma c_n^2 \leq \|f\|^2.$$

This is known as *Bessel's Inequality* and holds whether or not the orthonormal set is complete. But for a complete orthonormal set, we actually have *Bessel's Equation*:

$$\Sigma c_n^2 = \|f\|^2.$$

This also follows from the proof of Theorem 3 because

$$\|g_n\|^2 = c_1^2 + \ldots + c_n^2$$

and

$$\|g_n\|^2 \to \|f\|^2$$

as

$$\|f\|^2 - \|g_n\|^2 = \|g_n - f\|^2 \to 0.$$

On the other hand, if $\{c_n\}$ is a sequence of real numbers for which Σc_n^2 is convergent, we may define

$$g_n = c_1 e_1 + \ldots + c_n e_n$$

and note as before that

$$\|g_m - g_n\|^2 \to 0 \quad \text{as} \quad m, n \to \infty.$$

By the completeness of L^2, $\{g_n\}$ converges to an element g of L^2. If $n \geq i$, the Fourier coefficient
$$c_i = (g_n, e_i).$$
Let $n \to \infty$ and deduce that
$$c_i = (g, e_i)$$
for all i. In other words, every sequence $\{c_n\}$, for which Σc_n^2 is convergent, occurs as the sequence of Fourier coefficients of some function in L^2.

This information may be summarised in terms of a new linear space. Denote by l^2 the set of all sequences $x = \{x_n\}$ of real numbers for which Σx_n^2 is convergent. If x, y are two such sequences and a, b are real numbers, then $ax+by$ is the sequence whose n-th term is $ax_n + by_n$. Everything we said about L^2 in §7.3 has an analogue for l^2. The inequality
$$|x_n y_n| \leq \tfrac{1}{2}(x_n^2 + y_n^2)$$
shows that
$$\Sigma x_n y_n$$
is absolutely convergent if $x, y \in l^2$. From this we deduce that l^2 is a *linear space*, and also that there is a *scalar product* $(x.y$ or$)$
$$(x, y) = \Sigma x_n y_n$$
for any x, y in l^2. The *norm* of x is
$$\|x\| = \sqrt{(x, x)} = \{\Sigma x_n^2\}^{\frac{1}{2}}$$
and all the geometrical properties of l^2 follow just as before. This is not surprising as the above discussion gives the following fundamental result [20].

Theorem 6 (Riesz–Fischer). *The normed linear spaces L^2 and l^2 are isomorphic.*

Proof. Choose a complete orthonormal set $\{e_1, e_2, \ldots\}$ in L^2 and let
$$c_n = (f, e_n) \quad \text{for} \quad n = 1, 2, \ldots.$$
Then
$$f \to \{c_n\}$$
is a one–one linear mapping of L^2 onto l^2 which preserves the norm (by Bessel's Equation).

In view of the identity
$$4(f, g) = \|f+g\|^2 - \|f-g\|^2$$
in L^2, and a similar identity in l^2, it also follows that the scalar product is preserved, i.e.
$$(f, g) = \Sigma c_n d_n,$$

where c_n, d_n are the Fourier coefficients of f, g, respectively. This equation is generally attributed to *Parseval*.

By contrast with the canonical isomorphism of L^2 onto its dual space (Theorem 7.4.2), the isomorphism constructed here is vitally dependent on the choice of a complete orthonormal set.

Theorem 6 is perhaps all the more surprising if we recall from the beginning of §7.2 our claim that these results hold for $L^2(I)$, where I is any interval in $\mathbf{R}^k$ (subject to the obvious condition $m(I) > 0$). The only points of difference in the exposition are in the construction of a complete orthonormal set for $L^2(I)$ as in Theorem 4 where one reinterprets the idea of a rational step function, and in the proof of infinite dimensionality as in Proposition 1(i), where the condition $m(I) > 0$ is all that is required. In the historic development of quantum mechanics the 'matrix theory' of Heisenberg and the 'wave theory' of Schrödinger were finally unified by the latter – the essential step in the argument, from the mathematical point of view, being Theorem 6 in the case of $L^2(\mathbf{R}^k)$. See J. von Neumann [26].

In $L^2[-1, 1]$ the Gram–Schmidt process applied to the sequence

$$1, x, x^2, \ldots$$

yields the *normalised Legendre polynomials*

$$\sqrt{(n+\tfrac{1}{2})}\, P_n(x) \quad (n = 0, 1, 2, \ldots),$$

where
$$P_n(x) = \frac{1}{2^n n!} D^n (x^2 - 1)^n$$

and D is written for the operator $\dfrac{d}{dx}$.

The polynomials $1, x, x^2, \ldots$ do not belong to $L^2[0, \infty)$, but we may certainly look for a sequence $\{L_n(x)\}$ of polynomials which satisfy the orthogonality relations

$$\int_0^\infty L_m(x)\, L_n(x)\, e^{-x} dx = 0 \tag{9}$$

for $m \neq n$. The *Laguerre polynomials*

$$L_n(x) = e^x D^n (x^n e^{-x})$$

satisfy these relations; they may be obtained from the sequence

$$1, x, x^2, \ldots$$

by the Gram–Schmidt process, either by using a 'weighted' scalar product as in (9), or equivalently, we may begin with the sequence

$$e^{-\tfrac{1}{2}x}, x e^{-\tfrac{1}{2}x}, x^2 e^{-\tfrac{1}{2}x}, \ldots$$

in $L^2[0, \infty)$; the *normalised Laguerre functions* are then

$$\frac{1}{n!} L_n(x) e^{-\frac{1}{2}x} \quad (n = 0, 1, 2, \ldots).$$

In almost the same way we may construct in $L^2(-\infty, \infty)$ the *normalised Hermite functions*

$$(2^n n! \sqrt{\pi})^{-\frac{1}{2}} H_n(x) e^{-\frac{1}{2}x^2} \quad (n = 0, 1, 2, \ldots),$$

where $\qquad H_n(x) = (-1)^n e^{x^2} D^n(e^{-x^2}).$

It is of considerable importance in mathematical physics that these orthonormal sets are all complete in their respective L^2 spaces. We shall not establish this, but in the next section we shall prove that the trigonometric functions form a complete orthogonal set in $L^2(-\pi, \pi)$.

Exercises

1. Let $\{f_1, \ldots, f_m\}$ be a basis of the linear space M. Show that any f in M may be expressed uniquely as
$$f = a_1 f_1 + \ldots + a_m f_m$$
with $a_1, \ldots, a_m$ in $\mathbf{R}$.

2. Let M be a finite dimensional linear space. Show that a basis of M is a maximal linearly independent subset of M and also a minimal spanning set for M.

3. Use the proof of Lemma 1 to establish the following result. Let $\{f_1, \ldots, f_m\}$ be a basis of the linear space M and let $g_1, \ldots, g_n$ be linearly independent elements of M. Then $n \leq m$. If $n < m$ then $g_1, \ldots, g_n$, together with a suitable selection of $m - n$ of the f's form a basis of M.

4. Let M, N be finite dimensional subspaces of a given linear space L. Show that $M + N = \{m + n : m \in M, n \in N\}$ and $M \cap N$ are finite dimensional subspaces of L. Use Ex. 3 to extend a basis of $M \cap N$ to a basis of $M + N$ and hence prove that

$$\dim(M+N) + \dim(M \cap N) = \dim M + \dim N.$$

5. Let $\{f_1, \ldots, f_m\}$ be a basis for the linear subspace M of L^2. Suppose that $\|f_i\| = 1$ for each i, and consider the mapping $\phi: \mathbf{R}^m \to \mathbf{R}$ defined by $\phi((x_1, \ldots, x_m)) = \|x_1 f_1 + \ldots + x_m f_m\|$. Show that ϕ is continuous. Restrict ϕ to the unit sphere in $\mathbf{R}^m$ and use Theorem 9 of the Appendix to show that there is a real number $a > 0$ such that

$$\phi(x) \geq a|x|$$

for all x in $\mathbf{R}^m$. Establish the inequalities

$$a|x-y| \leq \|(x_1 - y_1)f_1 + \ldots + (x_m - y_m)f_m\| \leq m|x-y|$$

(where the norms on the extremes refer to $\mathbf{R}^m$).

6. (i) Show that $\mathbf{R}^m$ is a complete normed linear space.

(ii) Deduce from Ex. 5 that a finite dimensional subspace of L^2 is complete.

(iii) Deduce from (ii) that a finite dimensional subspace of L^2 is closed.

7. Calculate the Legendre, Laguerre and Hermite polynomials P_n, L_n, H_n for $n = 0, 1, 2$ and verify the orthogonality conditions that they satisfy in their respective L^2 spaces.

8. Let $u_n(x) = (x^2 - 1)^n$ for $n = 1, 2, \ldots$ and $u_0(x) = 1$ $(x \in \mathbf{R})$. Show that $Q_n(x) = D^n u_n(x)$ is a polynomial of degree n. If $m < n$ show, by repeated integration by parts, that

$$\int_{-1}^{1} x^m Q_n(x)\, dx = 0.$$

Deduce that
$$\int_{-1}^{1} Q_m(x) Q_n(x)\, dx = 0$$

for $m \neq n$.

Show that
$$\int_{-1}^{1} Q_n^2(x)\, dx = (2n)! \int_{-1}^{1} (1-x)^n (1+x)^n\, dx$$
$$= \frac{(n!)^2 2^{2n}}{(n+\tfrac{1}{2})}.$$

Hence show that the Legendre functions $\sqrt{(n+\tfrac{1}{2})}\, P_n$ $(n = 0, 1, \ldots)$ defined on p. 200 form an orthonormal set in $L^2[-1, 1]$.

7.6 Classical Fourier series

In this section we shall consider briefly the question of pointwise convergence. Our main reason for doing so is the sheer enjoyment of some of the famous theorems which played a significant part in the early researches following the publication of Lebesgue's thesis [11]–[14]. As a bonus we are able to exploit a very simple result about pointwise convergence (Proposition 3) to prove the completeness of the trigonometric functions $\cos kx$ $(k \geq 0)$, $\sin kx$ $(k \geq 1)$ in $L^2(-\pi, \pi)$.

If the trigonometric series

$$\tfrac{1}{2}a_0 + \sum_{k=1}^{\infty} (a_k \cos kx + b_k \sin kx)$$

converges to $f(x)$ for all real x, then the function f so defined is periodic, with period 2π, i.e.
$$f(x + 2\pi) = f(x)$$

for all real x. It is usual to begin with a function $f: \mathbf{R} \to \mathbf{R}$ and look for a trigonometric series whose sum is $f(x)$ for all x. To have any hope of success we must obviously insist that f has period 2π. Suppose that an arbitrary function $g: \mathbf{R} \to \mathbf{R}$ is given. We may define a periodic function f by letting

$$f(x + 2k\pi) = g(x) \quad \text{if} \quad x \in (-\pi, \pi]$$

($k = 0, \pm 1, \pm 2, \ldots$). Fig. 68 shows the graph of f when $g(x) = x$.

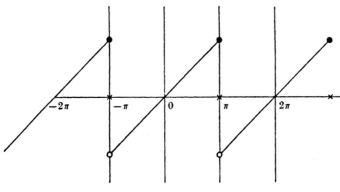

Fig. 68

Our choice of the *half*-open interval $(-\pi, \pi]$ facilitates the above definition of the periodic function f, but plays no role in the Fourier series expansion as the end points have measure zero. Incidentally,

$$L^2(-\pi, \pi] = L^2(-\pi, \pi)$$

and we shall in future use the latter notation as it is not quite so clumsy. As we shall see, the Fourier series for the illustrated function f converges to the 'average value', marked with a cross, at the doubtful points $(2k+1)\pi$.

We could equally well have chosen the interval $[0, 2\pi)$ but, of course, the periodic function f would then be quite different (Fig. 69).

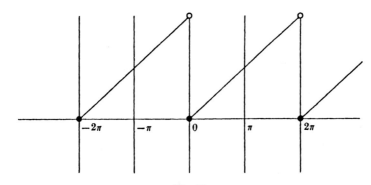

Fig. 69

In view of the finite length of the interval $(-\pi, \pi)$, the constant function 1 belongs to $L^2(-\pi, \pi)$ and so if $f \in L^2(-\pi, \pi)$ we deduce that $f1 = f \in L^1(-\pi, \pi)$. In other words

$$L^2(-\pi, \pi) \subset L^1(-\pi, \pi).$$

For any f in $L^1(-\pi, \pi)$ we may define the *Fourier series*

$$\tfrac{1}{2}a_0 + \sum_{k=1}^{\infty} (a_k \cos kx + b_k \sin kx),$$

where
$$a_k = \frac{1}{\pi} \int_{-\pi}^{\pi} f(t) \cos kt \, dt \quad (k \geq 0),$$
$$b_k = \frac{1}{\pi} \int_{-\pi}^{\pi} f(t) \sin kt \, dt \quad (k \geq 1).$$
(1)

These integrals exist because the (measurable) integrands are dominated by $|f|$ (Proposition 6.1.1).

One of the fundamental results of the whole theory is the **Riemann–Lebesgue Lemma** proved in §5.2.

Theorem 1. *If $f \in L^1(\mathbf{R})$ then the integrals*

$$\int_{-\infty}^{\infty} f(t) \cos kt \, dt, \quad \int_{-\infty}^{\infty} f(t) \sin kt \, dt$$

both exist and converge to 0 as $k \to \infty$.

In this theorem k is not restricted to the positive integers, but may take real values. As a particular case, if $f \in L^1(-\pi, \pi)$ we may replace f by $f\chi_{(-\pi,\pi)}$ in the Riemann–Lebesgue Lemma and deduce that the Fourier coefficients
$$a_k, b_k \to 0 \quad \text{as} \quad k \to \infty.$$

Now let $f \in L^1(-\pi, \pi)$ have period 2π and consider the partial sum

$$s_n(x) = \tfrac{1}{2}a_0 + \sum_{k=1}^{n} (a_k \cos kx + b_k \sin kx).$$

As the coefficients a_k, b_k are given by (1),

$$s_n(x) = \frac{1}{\pi} \int_{-\pi}^{\pi} f(t) \left\{ \frac{1}{2} + \sum_{k=1}^{n} (\cos kt \cos kx + \sin kt \sin kx) \right\} dt.$$

Recall the elementary trigonometric identity

$$\cos kt \cos kx + \sin kt \sin kx = \cos k(t-x).$$

Thus
$$s_n(x) = \frac{1}{\pi} \int_{-\pi}^{\pi} f(t) D_n(t-x) \, dt, \qquad (2)$$

where
$$D_n(x) = \frac{1}{2} + \sum_{k=1}^{n} \cos kx \qquad (3)$$

for all x in **R**. These functions D_n are called the *Dirichlet kernels*. By a simple translation
$$s_n(x) = \frac{1}{\pi} \int_{x-\pi}^{x+\pi} f(x+t) D_n(t) dt$$
$$= \frac{1}{\pi} \int_{-\pi}^{\pi} f(x+t) D_n(t) dt$$

because both f and D_n have period 2π. But
$$\int_{-\pi}^{0} f(x+t) D_n(t) dt = \int_{0}^{\pi} f(x-t) D_n(-t) dt,$$

and D_n is an even function, therefore
$$s_n(x) = \frac{1}{\pi} \int_0^{\pi} \{f(x+t) + f(x-t)\} D_n(t) dt. \qquad (4)$$

The question of pointwise convergence is now transformed to the following.

Proposition 1. *Let $f \in L^1(-\pi, \pi)$ have period 2π. Then $s_n(x) \to s(x)$ if and only if*
$$\int_0^{\pi} g(x,t) D_n(t) dt \to 0, \qquad (5)$$
where $g(x,t) = f(x+t) + f(x-t) - 2s(x).$

Proof. As a special case of (4), with $f = 1$, we deduce
$$1 = \frac{1}{\pi} \int_0^{\pi} 2 D_n(t) dt. \qquad (6)$$

(Or quite simply from definition (3).) If we multiply (6) by $s(x)$ and subtract from (4) the result is immediate.

There is another expression for $D_n(x)$ that is sometimes more convenient than (3). From the familiar identity
$$2 \sin \tfrac{1}{2}x \cos kx = \sin(k+\tfrac{1}{2})x - \sin(k-\tfrac{1}{2})x$$
we deduce that
$$(2 \sin \tfrac{1}{2}x) \sum_{k=1}^{n} \cos kx = \sum_{k=1}^{n} \{\sin(k+\tfrac{1}{2})x - \sin(k-\tfrac{1}{2})x\}$$
$$= \sin(n+\tfrac{1}{2})x - \sin \tfrac{1}{2}x,$$

which gives $(2\sin\tfrac{1}{2}x)D_n(x) = \sin(n+\tfrac{1}{2})x$

and so
$$D_n(x) = \frac{\sin(n+\tfrac{1}{2})x}{2\sin\tfrac{1}{2}x} \qquad (7)$$

provided x is not an integer multiple of 2π.

Many of the more delicate results on pointwise convergence of Fourier series are proved by a combination of the Riemann–Lebesgue Lemma with Proposition 1. For example:

Proposition 2 (*The Localisation Principle*). *Let $f \in L^1(-\pi, \pi)$ have period 2π. Then $s_n(x) \to s(x)$ if and only if, for some r in $(0, \pi]$,*
$$\int_0^r g(x, t) D_n(t)\, dt \to 0.$$

The notation is the same as for Proposition 1.

Proof. This is immediate from Proposition 1 as $(\sin\tfrac{1}{2}t)^{-1}$ is continuous and bounded on the interval $[r, \pi]$ and so
$$\int_r^\pi \frac{g(x, t)}{2\sin\tfrac{1}{2}t} \sin(n+\tfrac{1}{2})t\, dt \to 0$$

by the Riemann–Lebesgue Lemma. (If $r = \pi$ there is nothing to prove.)

This result is remarkable as it shows that the convergence or divergence of $\{s_n(x)\}$ is governed entirely by the behaviour of f in an arbitrarily small interval $(x-r, x+r)$; and this in spite of the fact that the Fourier series depends on the values of f throughout the interval $(-\pi, \pi)$. This principle can be applied, for example, if f is differentiable at the point x (see Ex. 14). As a very simple illustration of Proposition 2 we have:

Proposition 3. *Let ϕ be a step function and f the function of period 2π which coincides with ϕ on $(-\pi, \pi]$. Then the Fourier series of f at any point x converges to the sum*
$$\tfrac{1}{2}\{f(x+0)+f(x-0)\}. \qquad (\text{Fig. 70})$$

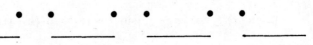

Fig. 70

For any $f: \mathbf{R} \to \mathbf{R}$ we denote by $f(x+0)$ the limit, if it exists, of $f(x+t)$ as t tends to 0 through positive values; similarly, $f(x-0)$ is the limit, if it exists, of $f(x-t)$ as t tends to 0 through positive values. This definition agrees with the one already given in §3.3 for $f(x+0)$, $f(x-0)$ when f is a monotone function (Ex. 1). The function f is continuous at x if and only if $f(x+0) = f(x-0) = f(x)$. These limits certainly exist for any step function and equally well for a periodic function f of the kind envisaged in Proposition 3.

Proof. Let
$$g(x,t) = f(x+t) + f(x-t) - f(x+0) - f(x-0).$$
For any given x, we can find $r > 0$ such that
$$f(x+t) - f(x+0) = 0$$
and also
$$f(x-t) - f(x-0) = 0$$
for $0 < t < r$. Thus $g(x,t) = 0$ for $0 < t < r$ and the Localisation Principle applies trivially.

Despite the apparent simplicity of this result and the comparative ease of the proof, the convergence of $s_n(x)$ near a point of discontinuity of f is remarkably subtle – see Ex. 19 for an illustration of *Gibbs' Phenomenon*. This simple result about pointwise convergence is also a key to our promised main theorem on strong convergence in $L^2(-\pi, \pi)$.

Theorem 2. *The trigonometric functions* $\cos kx$ $(k \geq 0)$, $\sin kx$ $(k \geq 1)$ *form a complete orthogonal set in* $L^2(-\pi, \pi)$.

Proof. (i) Let ϕ be a step function which vanishes outside $(-\pi, \pi)$ and let s_n be the n-th partial sum of the Fourier series of ϕ. From Proposition 3 it follows that $s_n \to \phi$ almost everywhere in $(-\pi, \pi)$. By Theorem 7.5.3 $\{s_n\}$ converges strongly to an element g of $L^2(-\pi, \pi)$, and by Theorem 7.2.3, a subsequence of $\{s_n\}$ converges pointwise almost everywhere to g. This now shows that $g = \phi$ almost everywhere and $\{s_n\}$ converges strongly to ϕ. In other words, given any $\epsilon > 0$, we can find a trigonometric polynomial s_n such that $\|\phi - s_n\| < \epsilon$.

(ii) Let $f \in L^2(-\pi, \pi)$. By part (i) of the proof of Theorem 7.5.4 we may find a step function ϕ vanishing outside $(-\pi, \pi)$ such that $\|f - \phi\| < \epsilon$. Combining with part (i) above we get $\|f - s_n\| < 2\epsilon$. Thus the trigonometric polynomials are dense in $L^2(-\pi, \pi)$, which is the result we set out to prove.

As we said in the more general setting of Theorem 7.5.5, this shows that the Fourier series of *any* f in $L^2(-\pi, \pi)$ converges *strongly* to f.

We now return to the question of pointwise convergence and extend Proposition 3 to a celebrated theorem of Jordan in which the step function ϕ is replaced by a function of *bounded variation* (see §3.5). To prepare for this we must first take a closer look at the Dirichlet kernels D_n (see Fig. 71). The discussion is so elementary that some of the details are set as exercises.

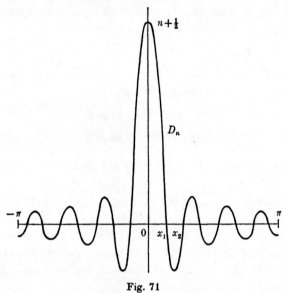

Fig. 71

Recall that
$$D_n(x) = \frac{\sin(n+\tfrac{1}{2})x}{2\sin\tfrac{1}{2}x}$$

for $x \neq 2k\pi$. Let us choose one fixed integer $n \geq 1$ and consider the integral
$$G(x) = \int_0^x D_n(t)\,dt$$

for $x \geq 0$. As D_n is continuous, $G'(x) = D_n(x)$ for all x (Theorem 3.4.1). By elementary calculus, G has alternate (local) maxima and minima at the points x, where $D_n(x) = 0$, viz. at the points
$$x_k = k\pi/(n+\tfrac{1}{2})$$

for $k = 1, 2, \ldots$. Let $x_0 = 0$. We may verify (Ex. 17) that the areas
$$|G(x_k) - G(x_{k-1})|$$

(between the graph of D_n and the intervals $[x_{k-1}, x_k]$) decrease as k increases from 1 to $n+1$, and hence that
$$0 \leq G(x) \leq G(x_1)$$

for all x in $[0, \pi]$. It follows that

$$|G(b) - G(a)| \leq G(x_1) \tag{8}$$

for all a, b in $[0, \pi]$.

Now let us take into account the integer n and denote by M_n the bound $G(x_1)$, viz.

$$M_n = \int_0^{\pi/(n+\frac{1}{2})} D_n(t)\, dt$$

which may also be written

$$M_n = \int_0^\pi \frac{\sin t}{t} \frac{(t/2n+1)}{\sin(t/2n+1)}\, dt.$$

These bounds have some interest in themselves (see Exx. 18, 19). For our present purpose it is sufficient to note that M_n decreases as n increases (Ex. 13). Hence (8) gives

$$\left|\int_a^b D_n(t)\right| \leq M_1 \tag{9}$$

for all $n = 1, 2, \ldots$, and all a, b in $[0, \pi]$. The constant

$$M_1 = \int_0^{\frac{1}{3}\pi} \frac{\sin 3t}{\sin t}\, dt = \tfrac{1}{2}\sqrt{3} + \tfrac{1}{3}\pi$$

is approximately 1·91, and so inequality (9) also holds for $n = 0$ as $D_0(t) = \tfrac{1}{2}$. Let us replace M_1 by the slightly more generous (and much more easily remembered) bound 2.

Lemma 1.
$$\left|\int_a^b D_n(t)\, dt\right| < 2$$

for all $n \geq 0$ and all a, b in $[0, \pi]$.

We are now in a position to prove

Theorem 3 (*Jordan*). *Let $f \in L^1(-\pi, \pi)$ have period 2π. If f has bounded variation on an interval $[x-r, x+r]$, where $0 < r \leq \pi$, then the Fourier series of f converges at the point x to the sum*

$$\tfrac{1}{2}\{f(x+0) + f(x-0)\}.$$

As f has bounded variation on $[x-r, x+r]$ we may express f as the difference of two increasing functions on that interval and so the limits $f(x+0), f(x-0)$ certainly exist (Ex. 1).

Proof. For the particular point x in question let

$$g(t) = f(x+t) + f(x-t) - f(x+0) - f(x-0)$$

for all t in $[0,r]$. Thus g has bounded variation on $[0,r]$ and $g(t) \to 0$ as $t \to 0$ through positive values. We may therefore express g as the difference $g_1 - g_2$ of two increasing functions on $[0,r]$ and (by subtracting a suitable constant from g_1, g_2) arrange that $g_1(t) \to 0$ and $g_2(t) \to 0$ as $t \to 0$ through positive values.

To any $\epsilon > 0$, there is a δ in $(0, r]$ such that
$$0 \leqslant g_i(t) < \epsilon \quad (i = 1, 2)$$
for all t in $(0, \delta)$. Now by Bonnet's form of the Mean Value Theorem (Ex. 6.2.12),
$$\int_0^\delta g_i(t) D_n(t)\, dt = g_i(\delta - 0) \int_{\delta_i}^\delta D_n(t)\, dt$$
for some δ_i in $[0, \delta]$. Thus by Lemma 1,
$$\left| \int_0^\delta g_i(t) D_n(t)\, dt \right| < 2\epsilon,$$
and as $g = g_1 - g_2$,
$$\left| \int_0^\delta g(t) D_n(t)\, dt \right| < 4\epsilon.$$

Having chosen δ we may apply the Riemann–Lebesgue Lemma to find an integer N such that
$$\left| \int_\delta^r g(t) D_n(t)\, dt \right| < \epsilon$$
for $n \geqslant N$. Finally
$$\left| \int_0^r g(t) D_n(t)\, dt \right| < 5\epsilon$$
for $n \geqslant N$, and this establishes Jordan's Theorem by the Localisation Principle.

As a very simple illustration of this result let us revert to the example mentioned at the beginning of the section where
$$f(x) = x \quad \text{for} \quad -\pi < x \leqslant \pi.$$
Now
$$\int_{-\pi}^\pi t \cos nt\, dt = 0 \quad (n \geqslant 0)$$
because the integrand is an odd function, and so all the a's are zero. Also,
$$\int_{-\pi}^\pi t \sin nt\, dt = \left[-t \frac{\cos nt}{n} \right]_{-\pi}^\pi + \int_{-\pi}^\pi \frac{\cos nt}{n}\, dt = (-1)^{n-1} 2\pi/n.$$
The Fourier series of $f(x)$ is therefore
$$2\{\sin x - \tfrac{1}{2} \sin 2x + \tfrac{1}{3} \sin 3x - \ldots\}.$$

As we claimed when we drew the graph of f, this converges to the average value 0 at the points of discontinuity $x = \pm\pi, \pm 3\pi, \ldots$. Substituting $x = \frac{1}{2}\pi$ gives the famous formula

$$\tfrac{1}{4}\pi = 1 - \tfrac{1}{3} + \tfrac{1}{5} - \tfrac{1}{7} + \ldots.$$

Another famous formula follows from Bessel's Equation (taking note that the trigonometric functions have to be normalised):

$$\frac{1}{\pi}\int_{-\pi}^{\pi} \frac{x^2}{4} dx = \Sigma \frac{1}{n^2},$$

i.e. $\qquad \tfrac{1}{6}\pi^2 = 1/1^2 + 1/2^2 + 1/3^2 + \ldots.$

There is another method of summing Fourier series by an averaging process. The general method is due to Cesàro but it was applied with remarkable success to Fourier series by Fejér. To get a rough idea of the process consider the series

$$1 - 1 + 1 - 1 + \ldots$$

which certainly does not converge: the sequence of partial sums is

$$1, 0, 1, 0, 1, \ldots.$$

But the sequence of 'running averages' is

$$1, \tfrac{1}{2}, \tfrac{2}{3}, \tfrac{1}{2}, \tfrac{3}{5}, \tfrac{1}{2}, \ldots$$

which is easily seen to converge to $\tfrac{1}{2}$. The averaging process smooths out to some extent the oscillation of the given series.

As before let $s_n(x)$ denote the partial sum

$$\tfrac{1}{2}a_0 + \sum_{k=1}^{n} (a_k \cos kx + b_k \sin kx)$$

and let $\qquad \sigma_n(x) = \dfrac{s_0(x) + s_1(x) + \ldots + s_n(x)}{n+1}$

for $n = 0, 1, 2, \ldots$. If $\{\sigma_n(x)\}$ converges to $s(x)$ we say that the Fourier series is *summable* $(C, 1)$ to $s(x)$. (Here C stands for Cesàro and 1 appears because there are (C, n) methods in which averages are taken n times.) As we shall see, there is much to be gained by this method, but the first simple point we must check is that nothing, in the way of convergence, is lost.

Proposition 4. *If* $s_n(x) \to s(x)$ *then* $\sigma_n(x) \to s(x)$.

Proof. All convergence is pointwise and for convenience in the proof we omit further reference to x.

First of all suppose that $s_n \to 0$. Given any $\epsilon > 0$, we may find N such that
$$|s_n| < \epsilon \quad \text{for} \quad n > N.$$
Thus
$$|\sigma_n| = \frac{|s_0 + \ldots + s_N + s_{N+1} + \ldots + s_n|}{n+1}$$
$$< \frac{|s_0 + \ldots + s_N|}{n+1} + \epsilon$$

for $n > N$. Having chosen such an N we may now find $N' > N$ so large that
$$\frac{|s_0 + \ldots + s_N|}{n+1} < \epsilon$$
for $n > N'$. Hence
$$|\sigma_n| < 2\epsilon$$
for $n > N'$. In other words $\sigma_n \to 0$.

If now $s_n \to s$ we may apply the above argument to the sequence $\{s_n - s\}$ for which the corresponding average sequence is $\{\sigma_n - s\}$. As $s_n - s \to 0$ it follows that $\sigma_n - s \to 0$, and this completes the proof.

The averaging process applied to equation (2) gives
$$\sigma_n(x) = \frac{1}{\pi} \int_{-\pi}^{\pi} f(t) F_n(t-x) \, dt, \tag{10}$$
where
$$F_n(x) = \frac{D_0(x) + D_1(x) + \ldots + D_n(x)}{n+1} \tag{11}$$

for $n = 0, 1, 2, \ldots$ defines the *Fejér kernels* F_n. As these functions F_n have period 2π, exactly the same proof as for Proposition 1 gives:

Proposition 5. *Let $f \in L^1(-\pi, \pi)$ have period 2π. Then $\sigma_n(x) \to s(x)$ if and only if*
$$\int_0^\pi g(x,t) F_n(t) \, dt \to 0, \tag{12}$$
where
$$g(x,t) = f(x+t) + f(x-t) - 2s(x).$$

To make use of F_n we need another explicit formula which is the analogue of formula (7) for D_n. The identity
$$2 \sin \tfrac{1}{2}x \sin (k + \tfrac{1}{2})x = \cos kx - \cos(k+1)x$$
may be summed for $k = 0, 1, \ldots, n$ and combined with (7) to give
$$(2 \sin \tfrac{1}{2}x)^2 F_n(x) = \frac{1 - \cos(n+1)x}{n+1}.$$

Thus
$$F_n(x) = \frac{1 - \cos(n+1)x}{4(n+1)(\sin \frac{1}{2}x)^2} \tag{13}$$

provided x is not an integer multiple of 2π.

These Fejér kernels are better behaved than the Dirichlet kernels – one vital property from our present point of view is that *the Fejér kernels are positive* (see Fig. 72).

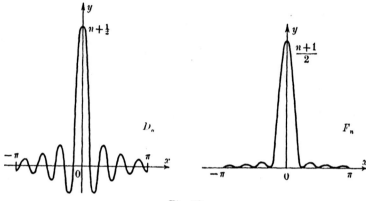

Fig. 72

Note also that (6) averages to give
$$1 = \frac{1}{\pi} \int_0^\pi 2F_n(t)\,dt,$$

i.e.
$$\int_0^\pi F_n(t)\,dt = \tfrac{1}{2}\pi \tag{14}$$

so the area under the graph of F_n (from $-\pi$ to π) has the constant value π, though the peak over the origin becomes higher as n increases. If $0 < \delta < \pi$ then
$$|F_n(x)| \leq 1/2(n+1)(\sin \tfrac{1}{2}\delta)^2 \tag{15}$$

for $\delta \leq x \leq \pi$, and this converges (uniformly) to zero as $n \to \infty$. We may now prove a famous theorem due to Fejér [7].

Theorem 4. *Let $f \in L^1(-\pi, \pi)$ have period 2π. For any point x at which the limits $f(x+0)$, $f(x-0)$ exist, the Fourier series of f is summable $(C, 1)$ to*
$$\tfrac{1}{2}\{f(x+0) + f(x-0)\}.$$

Proof. As before, let
$$g(x, t) = f(x+t) + f(x-t) - f(x+0) - f(x-0)$$

for the given point x. To any $\epsilon > 0$, there exists a δ in $(0, \pi)$ such that
$$|f(x+t)-f(x+0)| < \epsilon, \quad |f(x-t)-f(x-0)| < \epsilon \tag{16}$$
for $0 < t < \delta$. To apply Proposition 5 we split the interval of integration. In view of (14)
$$\left| \int_0^\delta g(x,t) F_n(t)\, dt \right| < 2\epsilon \int_0^\delta F_n(t)\, dt < \pi\epsilon.$$
Also, by (15),
$$\left| \int_\delta^\pi g(x,t) F_n(t)\, dt \right| \leq \int_\delta^\pi |g(x,t)|\, dt / 2(n+1)(\sin \tfrac{1}{2}\delta)^2.$$
Having chosen δ we may find N so large that this last expression is less than ϵ for $n \geq N$. Thus
$$\left| \int_0^\pi g(x,t) F_n(t)\, dt \right| < (\pi+1)\epsilon$$
for $n \geq N$. This gives the result by Proposition 5.

In the light of Proposition 4, Fejér's Theorem may be expressed in the eminently practical form: *If $f \in L^1(-\pi, \pi)$ has period 2π then*
$$\tfrac{1}{2}a_0 + \sum_{k=1}^{\infty} (a_k \cos kx + b_k \sin kx) = \tfrac{1}{2}\{f(x+0) + f(x-0)\}$$
provided both sides of this equation make sense, i.e. the series on the left hand side is convergent and the limits on the right hand side exist.

Fourier series were studied throughout most of the 19th century (from 1807 onwards) with a certain air of optimism that they would 'work' for any continuous function. But in 1876 du Bois-Reymond constructed a continuous function of period 2π whose Fourier series did not represent the function at a given point [6]. Fejér's Theorem (1904) is all the more remarkable as it shows that the Fourier series of a continuous function f of period 2π is summable $(C, 1)$ at every point x to the value $f(x)$. For the deduction from Fejér's Theorem of the Weierstrass Approximation Theorem, also the completeness of the trigonometric functions and the Legendre functions, the reader is invited to work through Exx. A–H.

Exercises

1. If $f: \mathbf{R} \to \mathbf{R}$ is a monotone function show that $f(x+0)$ and $f(x-0)$, as described in §3.3, are, respectively, the limits of $f(x+t)$ and $f(x-t)$ as t tends to 0 through positive values.

2. Let $f \in L^1(-\pi, \pi)$ have period 2π. If f is even, i.e. $f(-x) = f(x)$ for all x, show that the Fourier series of f has no sine terms. If f is odd, i.e.
$$f(-x) = -f(x)$$
for all x, show that the Fourier series of f has no cosine or constant terms.

3. Show that

(i) $\dfrac{\pi}{4} = \sum_{n=1}^{\infty} \dfrac{\sin(2n-1)x}{(2n-1)}$ if $0 < x < \pi$,

(ii) $x = \dfrac{\pi}{2} - \dfrac{4}{\pi} \sum_{n=1}^{\infty} \dfrac{\cos(2n-1)x}{(2n-1)^2}$ if $0 \leq x \leq \pi$.

4. Show that

(i) $x = 2 \sum_{n=1}^{\infty} \dfrac{(-1)^{n-1} \sin nx}{n}$ if $-\pi < x < \pi$,

(ii) $x^2 = \dfrac{\pi^2}{3} + 4 \sum_{n=1}^{\infty} \dfrac{(-1)^n \cos nx}{n^2}$ if $-\pi \leq x \leq \pi$.

5. (i) Let I be a subinterval of $(-\pi, \pi]$ and h_n the n-th partial sum of the Fourier series of χ_I. Use Lemma 1 to show that $h_n(t)$ is bounded for all n and all t.

(ii) Let $f \in L^1(-\pi, \pi)$ have period 2π and Fourier coefficients a_k, b_k given by equations (1). Show that
$$F(x) = \int_0^x f(t)\,dt = \tfrac{1}{2}a_0 x + \sum_{k=1}^{\infty} \frac{1}{k}(a_k \sin kx + b_k(1 - \cos kx))$$
for all x in $\mathbf{R}$. In other words *the Fourier series of f may be integrated term by term*.

(iii) Show that $\Sigma b_k/k$ is convergent and has sum
$$\frac{1}{2\pi} \int_{-\pi}^{\pi} F(x)\,dx.$$
If this constant is denoted by C then
$$F(x) = C + \tfrac{1}{2}a_0 x + \sum_{k=1}^{\infty} \frac{1}{k}(a_k \sin kx - b_k \cos kx).$$

6. Use Ex. 5 (iii) to simplify the second parts of Exx. 3, 4.

7. Use an appropriate Fourier series in conjunction with Bessel's equations to show that
$$\Sigma \frac{1}{n^4} = \frac{\pi^4}{90} \quad \text{and} \quad \Sigma \frac{1}{n^6} = \frac{\pi^6}{945}.$$

8. In the case where $f \in L^2(-\pi, \pi)$ use Parseval's equation of the previous section to give another proof of Ex. 5 (ii).

9. Let $f \in L^1(-\pi,\pi)$ have period 2π. Recall from §3.5 that the indefinite integral F of f is continuous and of bounded variation on any interval $[a,b]$. Let $G(x) = F(x) - \tfrac{1}{2}a_0 x$ for all x and show that G has period 2π. Use Jordan's Theorem to give another proof of Ex. 5(iii).

10. Use Ex. 5(i) to establish part (i) of the proof of Theorem 2 without referring to Theorems 7.5.3, 7.2.3.

11. Let
$$u(x) = \frac{1}{2\sin\tfrac{1}{2}x} - \frac{1}{x} \quad \text{if } 0 < x \leq \pi,$$
$$= 0 \quad \text{if } x = 0.$$

Show that u is a continuous function on $[0,\pi]$.

12. Let u be the function of Ex. 11.

(i) Apply the Riemann–Lebesgue Lemma to show that
$$\int_0^\pi u(t) \sin(n+\tfrac{1}{2})t\, dt \to 0$$
as $n \to \infty$.

(ii) Deduce from equations (6), (7) that
$$\int_0^{(n+\tfrac{1}{2})\pi} \frac{\sin x}{x} dx \to \tfrac{1}{2}\pi$$
as $n \to \infty$.

(iii) Hence show that
$$\int_0^X \frac{\sin x}{x} dx \to \tfrac{1}{2}\pi$$
as X tends to ∞ through real values.

13. Let
$$v(x) = \frac{\sin x}{x} \quad \text{if } x \neq 0,$$
$$= 1 \quad \text{if } x = 0.$$

Show that v is a continuous function on $\mathbf{R}$ and that v is strictly decreasing on the interval $[0,\pi]$.

14. Let $f \in L^1(-\pi,\pi)$ have period 2π and suppose that f is differentiable at a particular point x.

(i) Find r in $(0,\pi]$ such that
$$|g(x,t)/t| \leq 2(1 + |f'(x)|)$$
for $0 < t < r$. (Here g is the function described in Proposition 1 with $s(x) = f(x)$.)

(ii) Deduce from Ex. 11 or Ex. 13 that
$$\int_0^r \{g(x,t)/\sin\tfrac{1}{2}t\}\, dt$$
exists.

(iii) Prove that the Fourier series of f converges to $f(x)$ at the point x.

15. (i) Draw a rough sketch of the function v of Ex. 13.

(ii) Let
$$G(x) = \int_0^x \frac{\sin t}{t} dt$$
for $x \geq 0$. Show that G has alternate maxima and minima at the points $k\pi$ ($k = 1, 2, \ldots$).

(iii) Evaluate the areas between the graph of v and the intervals $[(k-1)\pi, k\pi]$ and show that they decrease to 0 as $k \to \infty$.

(iv) Express $G(n\pi)$ as an alternating sum and hence show that $G(n\pi)$ tends to a limit as $n \to \infty$. (We already know from Ex. 12 that this limit is $\frac{1}{2}\pi$.)

(v) Prove that
$$\left| \int_a^b \frac{\sin t}{t} dt \right| \leq G(\pi)$$
for all positive a, b.

16. Use Bonnet's form of the Mean Value Theorem (Ex. 6.2.12) to show that
$$\int_a^b \frac{\sin t}{t} dt$$
is bounded for all a, b in $\mathbf{R}$. (The previous example gives a least upper bound, viz.
$$\int_{-\pi}^{\pi} \frac{\sin t}{t} dt.)$$

17. Evaluate the area $(-1)^{k-1} \int_{x_{k-1}}^{x_k} D_n(t) dt$
used in the proof of Lemma 1 and show that it decreases as k increases from 1 to $n+1$. Hence prove that
$$0 \leq G(x) \leq G(x_1)$$
for all x in $[0, \pi]$. (Recall that $x_k = k\pi/(n+\frac{1}{2})$.)

18. Use Ex. 13 to show that the bounds M_n used in proving Lemma 1 decrease with n to the limit
$$M = \int_0^{\pi} \frac{\sin x}{x} dx.$$
(For further information about M see Ex. 19.)

19. (i) Let ϕ be the step function $\chi_{(0,\pi)} - \chi_{(-\pi,0)}$ and let f be the function of period 2π that coincides with ϕ on $(-\pi, \pi]$. Draw the graph of f.

(ii) Show that
$$f(x) = \frac{4}{\pi} \sum_{n=1}^{\infty} \frac{\sin(2n-1)x}{(2n-1)}$$
for all x in $\mathbf{R}$.

(iii) Let $s_n(x)$ be the n-th partial sum in the above series. Show that
$$s_n(x) = \frac{2}{\pi} \int_0^x \frac{\sin 2nt}{\sin t} dt$$

(iv) Show that s_n has alternate local maxima and minima at the points $x_k = k\pi/2n$ ($k = 1, 2, \ldots, 2n-1$).

(v) Show that $s_n(\pi/2n)$ is a global maximum of $s_n(x)$ on the interval $[0, \pi]$.

(vi) Show that $s_n(\pi/2n)$ decreases with n to the limit

$$L = \frac{2}{\pi} \int_0^\pi \frac{\sin t}{t}\, dt.$$

This limit is known as *Gibbs' constant*.

(vii) Compare the graph of f with those of s_n for large n and note that the latter are close to a vertical segment of height $2L$ through the origin. As L is approximately $1 \cdot 179$ this vertical segment is significantly longer than the jump in the function f at 0. (This is *Gibbs' Phenomenon*: it is exhibited by any function of bounded variation at an isolated jump discontinuity.)

20. Let $f \in L^1(-\pi, \pi)$ have period 2π and suppose that

$$A \leqslant f(x) \leqslant B$$

for all x. Show that the Cesàro sums σ_n satisfy the inequality

$$A \leqslant \sigma_n(x) \leqslant B$$

for all n and all x. (This rules out Gibbs' Phenomenon for the Cesàro sums of the function f described in the previous exercise.)

The following exercises are intended for students who have already met the ideas of uniform continuity and uniform convergence.

A. Let f be a real valued continuous function on the bounded closed interval $[a, b]$ and let $\epsilon > 0$ be given. To each point c of $[a, b]$ there exists $\delta_c > 0$ such that $|f(x) - f(c)| < \tfrac{1}{2}\epsilon$ for all x in $[a, b]$ satisfying $|x - c| < 2\delta_c$. Apply the Heine–Borel Theorem (Appendix, Theorem 6) to the open intervals $I_c = (c - \delta_c, c + \delta_c)$ and deduce the existence of $\delta > 0$ such that

$$|f(x) - f(y)| < \epsilon \text{ for all } x, y \text{ in } [a, b] \text{ satisfying } |x - y| < \delta.$$

This property is known as the *uniform continuity* of f on $[a, b]$.

B. Let f be a real valued continuous function of period 2π. Adapt the proof of Fejér's Theorem, using Ex. A, to show that $\{\sigma_n\}$ converges *uniformly* to f, i.e. given any $\epsilon > 0$; there exists an integer N such that $|\sigma_n(x) - f(x)| < \epsilon$ for all x in **R** provided $n \geqslant N$.

C. Let g be a real valued continuous function on $[0, \pi]$. Show that there is a continuous even function $f: \mathbf{R} \to \mathbf{R}$ of period 2π which coincides with g on $[0, \pi]$. For given $\epsilon > 0$, apply Ex. B to find a trigonometric polynomial σ_n which satisfies $|\sigma_n(x) - g(x)| < \epsilon$ for all x in $[0, \pi]$.

D. Use Ex. C to prove **Weierstrass' Approximation Theorem**: *Let*

g be a real valued continuous function on the bounded closed interval $[a, b]$. Then, to any $\epsilon > 0$, there is a polynomial p such that

$$|g(x) - p(x)| < \epsilon$$

for all x in $[a, b]$.
(You may quote the fact that the power series for the sine and cosine functions are uniformly convergent on any bounded interval.)

E. Show that the continuous functions $f: [a, b] \to \mathbf{R}$ form a *dense* subset of $L^2[a, b]$ (with respect to the L^2 norm).

F. Let $C[a, b]$ denote the linear space of continuous functions $f: [a, b] \to \mathbf{R}$ with norm

$$\|f\| = \sup\{|f(x)| : x \in [a, b]\}.$$

If $f_n, f \in C[a, b]$ and $\|f_n - f\| \to 0$, show that

$$\int_a^b |f_n - f|^2 \to 0.$$

(In other words, for $C[a, b]$ uniform convergence implies convergence in mean.)

G. Deduce from Exx. E, F that the Legendre polynomials form a *complete* orthogonal set in $L^2[-1, 1]$. (The orthogonality relations are established in Ex. 7.5.8.)

H. Show that the continuous functions of period 2π form a dense subset of $C[-\pi, \pi]$. Deduce once again that the trigonometric functions form a *complete* orthogonal set in $L^2[-\pi, \pi]$.

7.7 Reflections on Hilbert space

Throughout this book we have been reluctant to give lists of axioms. For example, the general definition of a linear space has been avoided by the simple expedient to considering only the special linear spaces which consist of functions $f: S \to \mathbf{R}$, where S is a (more or less arbitrary) set and $\mathbf{R}$ is the field of real numbers. A linear space L for us has been a collection of such functions which is closed with respect to addition and multiplication by scalars:

$$f, g \in L, \quad a, b \in \mathbf{R} \Rightarrow af + bg \in L.$$

This is a reasonable procedure because so many examples of linear spaces that arise in practice are of this kind. Even Euclidean space $\mathbf{R}^k$ may be thought of in this way if we regard

$$x = (x_1, \ldots, x_k)$$

as a function on $S = \{1, 2, \ldots, k\}$ with real values $x(i)$ written x_i.

Likewise, the elements $x = \{x_n\}$ of the linear space l^2 studied in §7.5 are real valued functions on $S = \{1, 2, \ldots\}$ with $x(n)$ written x_n.

For those who regard this as a mild form of cheating we propose now to give a general definition of linear space, but our main reason for doing so is to show that the essentially geometric approach of this chapter to the spaces L^p is capable of very great generalisation.

Let F be a field (as defined in §1.3) and L a set, such that, to any two elements f, g in L, there is a sum $f+g$ in L, and to any two elements a in F, f in L, there is a product af in L. Then L is called a *linear space* (or *vector space*) over F if the following axioms hold (for all f, g, h in L and all a, b in F):

L 1. $f+g = g+f$,
L 2. $(f+g)+h = f+(g+h)$,
L 3.   there is a *zero* element 0 in L such that $f+0 = f$,
L 4.   there is an element $-f$ such that $f+(-f) = 0$,
L 5. $a(f+g) = af+ag$,
L 6. $(a+b)f = af+bf$,
L 7. $(ab)f = a(bf)$,
L 8. $1f = f$, where 1 is the identity element of F.

We shall not play the usual games with these axioms and deduce the elementary rules for manipulating linear expressions

$$a_1 f_1 + \ldots + a_m f_m.$$

Suffice it to say that $(-1)f$ equals $-f$ (minus f), and if 0 is the zero element of F, $0f$ turns out to be the element 0 in L 3.

It is a very simple exercise to check that these axioms hold for the linear space L of *all* functions $f: S \to \mathbf{R}$. Our previous examples have been *linear subspaces* of this one, i.e. subsets of L which are closed with respect to sums and products as defined above.

Let L be a linear space over $\mathbf{R}$. A *norm* on L is a mapping $\| \ \|: L \to \mathbf{R}$ which satisfies the rules:

N 1. $\|f\| \geq 0$,
N 2. $\|f\| = 0$ if and only if $f = 0$,
N 3. $\|af\| = |a|\|f\|$,
N 4. $\|f+g\| \leq \|f\| + \|g\|$,

for any f, g in L, a in $\mathbf{R}$. A linear space with such a norm will be called a *normed linear space*.

Any normed linear space L may be regarded as a metric space if we define a distance

$$d(f, g) = \|f-g\|$$

for any f, g in L. The complete normed linear spaces are of fundamental importance in modern functional analysis and are called *Banach spaces*. Among the most famous examples of Banach spaces are the spaces L^p (Theorem 7.2.2) and the corresponding spaces l^p (Ex. 1); also Euclidean space $\mathbf{R}^k$ (Ex. 2), and the space $C[0, 1]$ of all continuous real valued functions on the interval $[0, 1]$ with respect to the norm defined by
$$\|f\| = \sup_{0 \leqslant x \leqslant 1} |f(x)|$$
(Ex. 3).

Let L be a linear space over $\mathbf{R}$. A *scalar product* on L associates with any two elements, f, g of L a real number (f, g) and satisfies the rules:

SP 1. $(f, g) = (g, f)$,
SP 2. $(af + bg, h) = a(f, h) + b(g, h)$,
SP 3. $(f, f) \geqslant 0$,
SP 4. $(f, f) = 0$ if and only if $f = 0$,
$(f, g, h \in L, a, b \in \mathbf{R})$.

Such a scalar product allows us to define a norm
$$\|f\| = \sqrt{(f, f)}.$$
The only non-trivial verification is the triangle inequality N 4 which may be deduced from Schwarz' Inequality
$$|(f, g)| \leqslant \|f\| \|g\|$$
exactly as in § 6.3. Let L be a linear space over $\mathbf{R}$ with scalar product; if L is complete with respect to the corresponding norm we shall call L a *Hilbert space*. All the geometric results of §§ 7.3, 7.4 were palpably based on the four rules for scalar products and the completeness of L^2; they hold without restriction for an abstract Hilbert space.

The most famous examples of Hilbert space are L^2 and l^2 (Theorem 7.5.6) and also $\mathbf{R}^k$ (Ex. 4). In fact, these are virtually the only examples if we restrict L to have 'countable dimension'. As a complete orthonormal set is the same as a maximal orthonormal set (Ex. 5), the existence of a complete orthonormal set in the abstract case is proved by waving the hand in an appropriate fashion and uttering the mystic words, 'Zorn's Lemma'. In the case of uncountable dimension the results of § 7.5 go through with very little change, once one realises that the sums involved can only have a countable set of non-zero terms. Halmos [9] is the obvious choice for the interested reader.

Exercises

1. The set l^p consists of all sequences $x = \{x_n\}$ of real numbers for which $\Sigma |x_n|^p$ is convergent ($p \geqslant 1$). Show that l^p is a Banach space with respect to the norm given by
$$\|x\| = \{\Sigma |x_n|^p\}^{1/p}.$$

2. Show that Euclidean space $\mathbf{R}^k$ is a Banach space with respect to the Euclidean norm defined in §4.1.

3. Show that $C[0, 1]$ is a Banach space with respect to the norm defined by
$$\|f\| = \sup_{0 \leqslant x \leqslant 1} |f(x)|.$$

4. Show that Euclidean space $\mathbf{R}^k$ is a Hilbert space with respect to the scalar product defined in §6.3.

5. A maximal orthonormal set in L^2 is one which is not contained in a strictly larger orthonormal set. Show that an orthonormal set in L^2 is maximal if and only if it is complete.

APPENDIX

THE ELEMENTS OF TOPOLOGY

In these few pages we shall consider three of the most fundamental ideas of topology – *Continuity, Connectedness and Compactness*; they are so basic to an education in the subject that we may call them 'The Three C's'. These ideas are important in §6.3 and §6.4 where we study the 'geometry of measure', and they are used extensively in the second volume. The only result in the Appendix of vital importance to the first five chapters is Theorem 6 (Heine–Borel) which is needed to prove Lemma 3.2.1 – the main step in establishing the consistency of our definition of the Lebesgue integral. The proof of the Heine–Borel Theorem may be read independently of the rest of the Appendix.

Roughly speaking, a continuous mapping $f: X \to Y$ takes points of X which are near one another into points of Y which are near one another. The most familiar way of expressing 'nearness' is in terms of distance; let us begin by making this idea precise. A *metric space* is a non-empty set X together with a *distance function*, or *metric*, d which associates to each pair of points x, y of X a real number $d(x, y)$ satisfying the rules:

M 1. $d(x,y) = d(y,x)$,
M 2. $d(x,y) \geq 0$,
M 3. $d(x,y) = 0$ if and only if $x = y$,
M 4. $d(x,z) \leq d(x,y) + d(y,z)$,

for any x, y, z in X. Strictly, a metric space is an ordered pair (X, d), but we shall often refer to X as the space and only mention d explicitly if this is necessary to avoid confusion.

Example 1. The most familiar example of a metric space is Euclidean space $\mathbf{R}^k$ in which the distance d is defined as in §4.1:

$$d(x,y) = \{(x_1-y_1)^2 + \ldots + (x_k-y_k)^2\}^{\frac{1}{2}}.$$

The *triangle inequality* M 4 is established in Proposition 6.3.4, Corollary.

Example 2. If X is a non-empty subset of $\mathbf{R}^k$ then any metric d on $\mathbf{R}^k$ may be restricted to points x, y in X and so defines a metric on X.

Example 3. We may define another distance d on $\mathbf{R}^k$ by the equation

$$d(x,y) = \max\{|x_1-y_1|, \ldots, |x_k-y_k|\}$$

and verify M 1–M 4 immediately (cf. §6.4 in which this metric is used).

Example 4. For any non-empty set X we may define the *discrete metric*:
$$d(x,y) = 1 \quad \text{if} \quad x \neq y,$$
$$= 0 \quad \text{if} \quad x = y.$$

Example 5. Let $C[0,1]$ denote the set of all continuous functions $f:[0,1] \to \mathbf{R}$ (as defined in §3.3) and let
$$d(f,g) = \sup_{0 \leqslant x \leqslant 1} |f(x) - g(x)|.'$$
(Here we anticipate, as we did in §3.3, the fact that f and g are bounded: this is proved in Theorem 9 below.)

To begin with, nearness in the metric space (X, d) is expressed in terms of the *ball* $\quad B(a, r) = \{x \in X : d(a, x) < r\}$
($a \in X, r > 0$). The points 'near' a are the points which lie in $B(a, r)$ for some 'small' r. We naturally call a the *centre* and r the *radius* of the ball $B(a, r)$ and represent such a ball by a 'completely round' ball in $\mathbf{R}^2$ (Fig. A. 1). But it is just as well to realise from the beginning that this diagram may be misleading. For example, if X consists of the points $x = (x_1, x_2)$ in $\mathbf{R}^2$ with $x_1 \geqslant 0, x_2 \geqslant 0$ and the usual Euclidean distance, then the ball $B(0, r)$ in X would be as in Fig. A. 2.

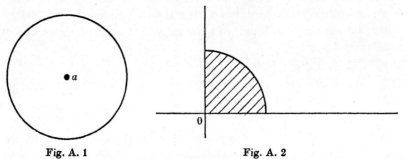

Fig. A. 1   Fig. A. 2

In Example 3 the ball $B(a, r)$ would be the interior of a square as in Fig. A. 3.

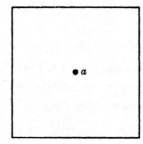

Fig. A. 3

In Example 4 the ball $B(a,r)$ consists of the single point a – the centre – if $0 < r \leqslant 1$, and is the whole space X if $1 < r$. In the last mentioned case, of course, we cannot recapture either the centre or the radius when the ball is given! In Example 5 the ball $B(g,r)$ consists of all continuous functions $f\colon [0,1] \to \mathbf{R}$ whose graphs lie in the indicated strip (Fig. A. 4).

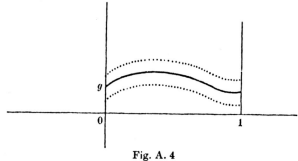

Fig. A. 4

The idea of a ball in a metric space is still very closely associated with the distance function. But one further move towards a truly geometric idea of nearness takes us into the heart of general topology. A subset G of X is *open* if, to any point g of G, there is a radius $r > 0$ such that $B(g,r) \subset G$ (Fig. A. 5). In rough and ready terms, an open set is one in which there is 'elbow room' around every point. It is immediately verified that the ball $B(a, \delta)$ is open: in the Fig. A. 6 we need only take $r = \delta - d(a,g)$.

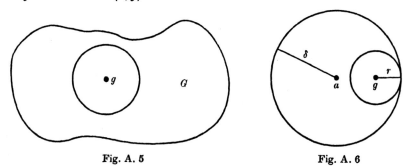

Fig. A. 5   Fig. A. 6

The empty set $\varnothing$ qualifies, in a rather negative way, as an open set, because it has no points around which to require elbow room. Also the whole space X is open as it possesses all the room there is. In dealing with the subsets of X it is natural to ask what happens to intersections and unions. As far as the open sets are concerned we have the following fundamental properties.

APPENDIX

Proposition 1. *In a metric space* (X, d):

O 1. $\varnothing, X$ *are open sets*,
O 2. *the union of any collection of open sets is open*,
O 3. *the intersection of any two open sets is open*.

Proof. (i) We have just verified O 1.

(ii) Let G be the union of the open sets $G_i (i \in I)$. If $a \in G$, then $a \in G_i$ for some i. Thus $B(a, r) \subset G_i$ for some $r > 0$ and *a fortiori* $B(a, r) \subset G$.

(iii) Let G_1, G_2 be open. If $a \in G_1 \cap G_2$ then $B(a, r_1) \subset G_1$ and $B(a, r_2) \subset G_2$ for suitably small $r_1, r_2 > 0$ (Fig. A. 7). Let $r = \min \{r_1, r_2\}$; then $B(a, r) \subset G_1 \cap G_2$.

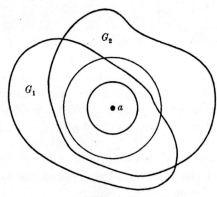

Fig. A. 7

Property O 3 extends at once to any finite collection of open sets, but is not true in general for an infinite collection of open sets. For example, in $\mathbf{R}^k$ the open balls $B(a, 1/n)$ $(n = 1, 2, ...)$ intersect in the set $\{a\}$ which is not open in $\mathbf{R}^k$.

We shall now develop a geometric point of view in which we express, wherever possible, each new idea in terms of open sets, without reference to distance, and use only the three properties of Proposition 1. As we shall see, this approach is remarkably rich in applications. (The experienced reader will recognise that we are in fact discussing general topological spaces, although, for the moment, we continue to refer to metric spaces in the results we prove.)

In the evolution of the subject it was not the open sets that came first; rather, the focus was upon limits, and sets which were closed with respect to limits. Let X be a metric space and A a non-empty subset of X. A *limit point* of A is a point p of X such that every open set containing p also contains at least one point of A distinct from p. A subset of X is *closed* if and only if it contains all its limit points. For example, the closed interval $[a, b]$ in **R** is indeed closed as its name would suggest, and every point of $[a, b]$ is a limit point. Every point of the half open interval $[0, 1)$ is a limit point, and in addition, 1 is a limit point. The points $1/n$ $(n = 1, 2, \ldots)$ form a set with just one limit point, viz. 0, which does not belong to the set. It turns out that closed sets are very easily described in terms of open sets.

Proposition 2. *A subset F of the metric space X is closed if and only if the complementary set $X \backslash F$ is open.*

Write F^c as shorthand for $X \backslash F$.

Proof. (i) Let F^c be open. Then any point of F^c fails, by our definition, to be a limit point of F. In other words, any limit point of F must lie in F, and so F is closed.

(ii) Assume that F^c is not open. Then there is a point p of F^c such that every open set containing p also contains points of F. In other words $p \in F^c$ and is a limit point of F. Thus F is not closed.

To illustrate the ideas of open and closed sets, consider an arbitrary subset A of the metric space X. Each point p of X falls into exactly one of the following three categories:

(i) there is an open set containing p and contained in A,
(ii) there is an open set containing p and contained in A^c,
(iii) every open set containing p has non-empty intersection with both A and A^c.

The points in category (i) form the *interior* of A, written int A, those in (ii) form the *exterior* of A, written ext A, and those in (iii) form the *boundary* of A, written ∂A. Thus X is the union of the disjoint sets int A, ext A, ∂A.

If G is an open set contained in A, any point p of G is, by definition, an interior point of A; moreover, any point p of int A is contained in such an open set G. It follows that int A is the union of all the open subsets of A. In view of property O 2, int A is an open subset of A; in fact int A *is the largest open subset of A*, as it contains all the others. By exactly the same argument ext A is the largest open subset of A^c. Rather more interestingly, by taking complements in X, we see that *there is a smallest closed set containing A* called the *closure* of A and written $\bar{A}$, viz., the complement of ext A. Thus $\bar{A}$ may be described as the intersection of all closed sets containing A, or equivalently as int $A \cup \partial A$. The reader is invited in Ex. 2 to show that the closure of a set is obtained by adjoining to the set all its limit points.

For most of the familiar sets in $\mathbf{R}^k$ these concepts of interior, exterior and boundary are just as one would expect. For example, if $A = B(a, r)$ then int $A = A$, ∂A is the *sphere* $S(a, r) = \{x \in X : d(a, x) = r\}$ and ext $A = \{x \in X : d(a, x) > r\}$. But if A consists of the points of $\mathbf{R}^k$ with rational co-ordinates, int A, ext A are both empty and ∂A is the whole of $\mathbf{R}^k$!

We are now in a position to discuss the three C's. First continuity. Let X, Y be metric spaces with corresponding metrics d, d', respectively. According to the classical definition, a mapping $f: X \to Y$ is *continuous at the point* a of X if, given any $\epsilon > 0$, there exists $\delta > 0$ such that

$$d'(f(x), f(a)) < \epsilon \quad \text{for all } x \text{ in } X \text{ satisfying} \quad d(x, a) < \delta. \tag{1}$$

If f is continuous at every point of X, we say that f is *continuous*. This analytic definition may be given a more geometric look if we set $b = f(a)$ and rewrite (1) as

$$f(B(a, \delta)) \subset B(b, \epsilon). \tag{2}$$

(For any subset A of X, the image set $f(A)$ consists of all $f(x)$ for x in A.) This criterion for continuity still depends on the idea of distance, but we now give a criterion entirely in terms of open sets.

Proposition 3. *Let X, Y be metric spaces. The mapping $f: X \to Y$ is continuous if and only if*

$$\text{for every open subset } G \text{ of } Y, f^{-1}(G) \text{ is an open subset of } X. \tag{3}$$

The 'inverse image' $f^{-1}(G)$ denotes the set of all x in X such that $f(x)$ lies in G.

Proof. (i) Assume that f is continuous. Let G be an open subset of Y, and let a be an arbitrary point of $f^{-1}(G)$. Thus $b = f(a)$ belongs to G and we may find an open ball $B(b, \epsilon)$ contained in G. As f is continuous we may find an open ball $B(a, \delta)$ which satisfies (2) and this implies that
$$B(a, \delta) \subset f^{-1}(G).$$
Thus $f^{-1}(G)$ is open.

(ii) Assume that condition (3) holds. Let a be an arbitrary point of X and $B(b, \epsilon)$ an arbitrary open ball, centre $b = f(a)$. The point a belongs to the open set $f^{-1}(B(b, \epsilon))$ which therefore contains an open ball $B(a, \delta)$ for some $\delta > 0$. This verifies condition (2) and shows that f is continuous at a.

It often occasions surprise that (3) features the *inverse* image $f^{-1}(G)$.

A constant mapping $f: \mathbf{R} \to \mathbf{R}$ is certainly continuous and carries all the open subsets of $\mathbf{R}$ (except the empty set) onto a single point of $\mathbf{R}$. Thus we cannot, in general, expect a continuous mapping $f: X \to Y$ to carry open subsets of X onto open subsets of Y.

Open sets and closed sets are complementary in the sense of Proposition 2. Nevertheless the half-open interval $[0, 1)$ in $\mathbf{R}$ is an example of a set which is neither open nor closed; also the empty set $\varnothing$ and the 'universal set' X are both open in X and hence both closed in X. In a discrete metric space (Example 4) every subset is open and closed (Ex. 8). At the other extreme a *connected space* X is one in which the only open and closed subsets are $\varnothing$, X. If A is an open and closed subset of X, then $B = X \backslash A$ is likewise an open and closed subset of X. Thus X is connected if and only if it is impossible to express X as the union of two non-empty disjoint open subsets. This agrees with the very natural idea that a connected set cannot be in two pieces.

Let us try to identify the connected subsets of the real line. Consider the set
$$E = [0, 1) \cup (2, 3]. \qquad \text{(Fig. A. 8)}$$

Fig. A. 8

Intuition tells us that E is not connected because E falls into two pieces $[0, 1), (2, 3]$. But neither of these pieces is an open subset of $\mathbf{R}$; so in what sense is E the union of two open sets? The answer is that E inherits a metric from $\mathbf{R}$ and the sets in question are open subsets of E for this inherited metric: in fact $[0, 1) = B_E(0, 1)$ and $(2, 3] = B_E(3, 1)$ in the inherited metric. Quite generally, if (X, d) is a metric space and E a non-empty subset of X, then the restriction d_E of d to pairs x, y in E defines a metric space (E, d_E). We say that E is a *connected* subset of X if (E, d_E) is a connected space as defined above. From the geometric point of view there is a better way of doing this.

Proposition 4. *The open subsets of the metric space (E, d_E) are precisely the subsets of the form $E \cap G$ where G is open in (X, d).*

Proof. (i) The set $E \cap G$ is obviously open in (E, d_E).

(ii) Let H be an open subset of E for the inherited metric. To any point x of H there is a radius r_x such that

$$E \cap B(x, r_x) \subset H.$$

Let G be the union of these open balls $B(x, r_x)$ in X as x varies in H. Then G is open (by property O 2) and $H = E \cap G$.

We call these sets $E \cap G$ *relatively open* in E. Thus E is a connected subset of X if and only if E cannot be expressed as the union of two non-empty disjoint relatively open subsets. In the above example, $[0, 1) = E \cap (-1, 1)$ and $(2, 3] = E \cap (2, 4)$ are relatively open subsets of E.

The following result seems obvious enough, but depends heavily on the completeness of **R**.

Theorem 1. *The connected subsets of* **R** *are precisely the intervals.*

Recall from §1.1 the nine types of interval on **R**:

$$[a,b], (a,b), [a,b), (a,b];$$
$$[a,\infty), (a,\infty), (-\infty,a], (-\infty,a);$$
$$(-\infty,\infty)$$

(including the empty interval (a, a)).

Proof. (i) Let I be a connected subset of **R**. If $a, b \in I$ and $a < c < b$, then $c \in I$ — otherwise the open subsets $(-\infty, c), (c, \infty)$ would cover I and 'disconnect' I (just as the open subsets $(-1, 1), (2, 4)$ served to 'disconnect' E in the above example). In view of this 'intermediate point' property we may show that I is an interval.

The result is obviously true if I is empty so assume from now on that I contains a point p.

Suppose that I is unbounded above. Then to any x satisfying $p < x$ there is an element y of I satisfying $p < x < y$ and so $x \in I$. Thus $[p, \infty) \subset I$ in this case.

On the other hand suppose that I is bounded above and let $b = \sup I$. If $p < x < b$, there is a point y in I such that $p < x < y$ and so $x \in I$; thus $[p, b) \subset I$ in this case. The point b may or may not belong to I, but $x \leqslant b$ for all x in I as b is an upper bound.

Similarly we may consider the cases where I is bounded or unbounded below. There are then exactly the nine cases listed after the statement of the theorem.

(ii) Assume that A, B are non-empty disjoint relatively open subsets of the closed interval $[a, b]$ such that $A \cup B = [a, b]$ and $a \in A, b \in B$. We shall show that this case cannot occur. Let $c = \sup A$; thus $a \leqslant c \leqslant b$. If $c \in A$, then $c < b$ (A, B are disjoint) and so there exists an interval $[c, c+h) \subset A$ ($h > 0$). This contradicts the fact that c is an upper bound of A. If $c \in B$, then $a < c$ and so there exists an interval $(c-h, c] \subset B$ ($h > 0$). This contradicts the fact that c is the *least* upper bound of A.

In general let I be an interval and assume that A, B are non-empty disjoint relatively open subsets whose union is I. Without loss of generality we may find a, b in I with $a < b$ such that $a \in A, b \in B$. The above argument now applies to the intersections of I, A, B with $[a, b]$ and yields a contradiction. Thus I is connected as we set out to prove.

We now ask what happens to a connected set if we apply a continuous mapping to it. The answer is remarkably simple.

Theorem 2. *Let f be a continuous mapping of X onto Y. If X is connected then so is Y.*

Proof. Suppose that $Y = C \cup D$, where C, D are non-empty disjoint open subsets of Y. Then

$$A = f^{-1}(C), \quad B = f^{-1}(D)$$

are non-empty disjoint sets whose union is X. Moreover A, B are open subsets of X by Proposition 3. Thus X is not connected.

To give some indication of the power of Theorem 2 let us note the following result which is fundamental in the theory of (polynomial) equations.

Theorem 3 (Intermediate Value Theorem). *Let f be a continuous real valued function on the interval $[a, b]$. Then f takes every value between $f(a)$ and $f(b)$.*

Proof. By Theorems 1, 2, $f([a, b])$ is an interval containing $f(a)$ and $f(b)$.

As an immediate consequence of Theorem 3 we also have:

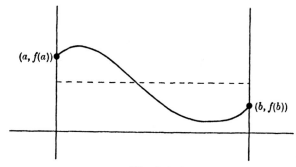

Fig. A. 9

Theorem 4 (Fixed Point Theorem). *Let I be a bounded closed interval on $\mathbf{R}$. If $g: I \to I$ is continuous then there is a point p of I such that*

$$g(p) = p.$$

Proof. Let $I = [a, b]$. If $g(a) = a$ or $g(b) = b$ there is nothing to prove. Therefore we may assume that $g(a) > a$ and $g(b) < b$. Let

$$f(x) = x - g(x) \quad (x \in I).$$

Then $f(a) < 0, f(b) > 0$ and so by Theorem 3 there is a point p of (a, b) such that $f(p) = 0$.

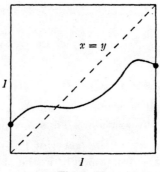

Fig. A. 10

From the point of view of the Figs. A. 9, 10 these last two results appear to be obvious. In each case the graph begins on one side of a given (dotted) line and ends on the other side. Alternative proofs may be given by proving that the graphs are connected (see Ex. 12). In the same vein there is a much more general result.

Theorem 5. *Let A be a subset of X. If E is a connected subset of X whose intersection with ∂A is empty, then either*

$$E \subset \operatorname{int} A \quad \text{or} \quad E \subset \operatorname{ext} A. \qquad \text{(Fig. A. 11)}$$

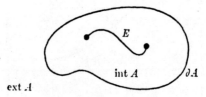

Fig. A. 11

ELEMENTS OF TOPOLOGY

Proof. Recall that the disjoint sets int A, ext A, ∂A cover X. As $E \cap \partial A$ is empty the relatively open sets

$$E \cap \text{int}\, A, \quad E \cap \text{ext}\, A$$

cover E and so one of them must be empty.

We made important use of this result in the proof of two main theorems on the transformation of integrals in §6.4 (specifically, in the proof of Proposition 6.4.1).

A subset E of $\mathbf{R}^k$ is *bounded* if E is contained in some suitably large ball $B(a, r)$. These bounded sets cannot be described purely in terms of open sets, without reference to the Euclidean metric (see Ex. 13), but it is of the utmost importance that the *bounded closed subsets* of $\mathbf{R}^k$ can be so described.

Theorem 6 (Heine–Borel). *Let C be a bounded closed subset of $\mathbf{R}^k$. In any collection of open sets which cover C there is a finite selection of open sets which also cover C.*

We say that the sets G_i $(i \in I)$ cover C if C is contained in their union $\bigcup G_i$.

Proof. For simplicity of notation and ease of illustration we shall write out the proof for the case of $\mathbf{R}^2$ – but the method generalises at once to $\mathbf{R}^k$ $(k \geqslant 1)$.

(i) Consider first the case of a closed square Q and a covering of Q by open sets G_i $(i \in I)$. Let us assume that there is no finite collection of these G's covering Q and look for a contradiction.

Divide Q into four equal closed squares by halving the sides. At least one of these cannot be covered by a finite collection of G's; otherwise Q could be so covered. Select one and call it Q_1. (For definiteness, a simple rule of selection may be given such as the one that favours left to right and then favours lower to upper.) Now divide Q_1 in the same way into four equal closed squares and select one, Q_2, which cannot be covered by a finite collection of G's. Continuing in this way by successive halving we construct a decreasing sequence of closed squares

$$Q_1 \supset Q_2 \supset Q_3 \supset \ldots$$

none of which can be covered by a finite collection of G's (Fig. A.12). Let (a_n, b_n) be the co-ordinates of the lower left hand corner of Q_n. Then the sequences $\{a_n\}$, $\{b_n\}$ are increasing and bounded, and so converge to limits a, b. For $n \geqslant m$ the point (a_n, b_n) lies in the closed square Q_m and so (a, b) lies in Q_m $(m = 1, 2, \ldots)$. We are now able to produce a contradiction. There is an open set G_i of our given covering

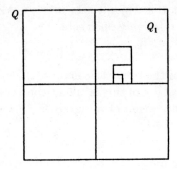

Fig. A. 12

which contains the point (a, b). For some $\delta > 0$ the open ball centre (a, b) radius δ lies in G_i. For sufficiently large m the diagonal of Q_m is less than δ and so Q_m lies entirely in G_i. Here is our contradiction, for Q_m is covered by just *one* set G_i (Fig. A. 13).

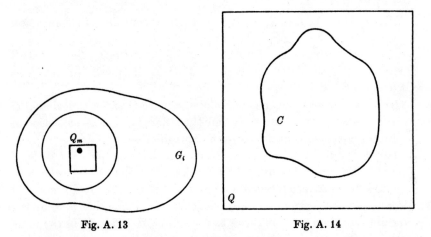

Fig. A. 13　　　　　　　　Fig. A. 14

(ii) Let C be a bounded closed subset of $\mathbf{R}^2$. As C is bounded we may find a closed square Q containing C (Fig. A. 14). Suppose that C is covered by the collection $\{G_i\}$ of open sets. Then Q (even $\mathbf{R}^2$) is covered by this same collection augmented by the single open set C^c ($C^c = \mathbf{R}^2 \setminus C$). According to part (i), Q may be covered by a finite selection of the G's and this extra open set C^c. But C^c does nothing to cover C and so C is covered by the finite selection of G's by themselves.

This famous theorem leads to a general definition. A (metric) space X is *compact* if, from any collection of open sets of X which cover

X, there is a finite selection of open sets which also cover X. More generally, let C be a subset of X. The open sets G_i in X cover C if and only if the sets $C \cap G_i$ cover C. Thus we say that C is a *compact* subset of X if C is a compact space in the sense of the above definition using the relatively open sets inherited from X. The Heine–Borel Theorem therefore asserts that any bounded closed subset of $\mathbf{R}^k$ is compact. The converse is much easier to establish and so we have the following characterisation of the compact subsets of $\mathbf{R}^k$.

Theorem 7. *The compact subsets of $\mathbf{R}^k$ are precisely the bounded closed subsets.*

Proof. It only remains to prove one half. Assume that C is compact.

(i) The open balls $B(0, n)$ $(n = 1, 2, \ldots)$ cover the whole space and so cover C. By the compactness, a finite selection of these will cover C, and as they are concentric, just one of them will cover C. Thus C is bounded.

(ii) Let a be a point not belonging to C. (If no such point exists then $C = \mathbf{R}^k$ is closed.) Consider the exteriors G_n of the balls

$$B(a, 1/n) \quad (n = 1, 2, \ldots).$$

Each point c of C is at a distance $d(c, a) > 0$ from a and so lies in G_n for sufficiently large n. Thus the open sets G_n cover C. By the compactness a finite selection of the G_n's will cover C and as they are concentric, $G_m \supset C$ for some $m \geq 1$. In other words, all points x satisfying

$$d(x, a) \leq 1/m$$

lie in the complement of C. This complement is therefore open and C is closed.

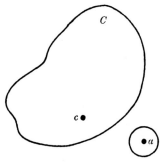

Fig. A. 15

There is a general topological theorem which is the analogue for compactness of Theorem 2. Here again the proof is almost immediate from the definitions.

Theorem 8. *Let X be compact and f a continuous mapping of X onto Y. Then Y is compact.*

Proof. Let the open sets H_i cover Y. Then the open sets

$$G_i = f^{-1}(H_i)$$

cover X. From these G_i's we select $G_{i_1}, G_{i_2}, \ldots, G_{i_n}$ which cover X and then $H_{i_1}, H_{i_2}, \ldots, H_{i_n}$ cover Y.

As an immediate application we have:

Theorem 9. *Let f be a continuous real valued mapping on a compact (metric) space X. Then the values of f are bounded and f attains the values*

$$\sup_{x \in X} f(x), \quad \inf_{x \in X} f(x).$$

Proof. The image $f(X)$ is a compact subset of **R** and so is bounded and closed by Theorem 7. Let

$$s = \sup_{x \in X} f(x);$$

then to any $\epsilon > 0$, there exists x in X such that $s - \epsilon < f(x) \leqslant s$. Either $s = f(x)$ for one of these points x, or s is a limit point of the closed set $f(X)$; in each case $s \in f(X)$. In the familiar classical language, the maximum value s is attained.

The argument with inf is almost identical.

There are two famous results from the calculus which follow from Theorem 9.

Theorem 10 (Rolle's Theorem). *Let the continuous function $f: [a, b] \to \mathbf{R}$ be differentiable at every point of the open interval (a, b). If $f(a) = f(b)$ then there is a point c in (a, b) such that $f'(c) = 0$.*

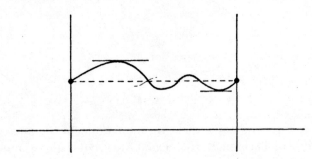

Fig. A. 16

Proof. If f is constant the theorem is obvious. In any other case we can find a point c in (a,b) at which $f(c)$ is either the (global) maximum or minimum value attained by f on $[a,b]$. Since f is differentiable at c

$$\frac{1}{h}\{f(c+h)-f(c)\} \to f'(c)$$

as $h \to 0$. If $f'(c) > 0$ or $f'(c) < 0$, then $f(c+h) - f(c)$ changes sign as h increases through the value 0. This contradicts the maximum or minimum property and proves that $f'(c) = 0$.

Theorem 11 (Mean Value Theorem). *Let the continuous function*

$$f:[a,b] \to \mathbf{R}$$

be differentiable at every point of the open interval (a,b). Then there is a point c in (a,b) such that

$$f(b)-f(a) = (b-a)f'(c).$$

Proof. Let $g(x) = f(x) - kx$ and choose k so that $g(a) = g(b)$, i.e. so that

$$f(b)-f(a) = k(b-a).$$

As g now satisfies the conditions of Rolle's Theorem there is a point c in (a,b) such that $g'(c) = 0$, i.e. such that

$$f'(c) = k.$$

This completes the proof.

Another application of Theorem 9 to the problem of eigenvalues is given in Ex. 16.

Although we have restricted our attention to metric spaces, many of the above ideas are capable of wide generalisation. We mention very briefly how this may be done. A *topological space* is a non-empty set X together with a collection $\mathscr{T}$ of subsets of X (which we call *open* subsets) satisfying the following conditions:

O 1. $\varnothing, X \in \mathscr{T}$,
O 2.   if $G_i \in \mathscr{T}$ for all i in I, then $\bigcup G_i \in \mathscr{T}$,
O 3.   if $G_1, G_2 \in \mathscr{T}$, then $G_1 \cap G_2 \in \mathscr{T}$.

Strictly speaking, a topological space is an ordered pair $(X, \mathscr{T})$, but it is usual to refer to X as the (topological) space, and, if necessary, refer to $\mathscr{T}$ as the *topology* on X. In view of Proposition 1 a metric space (X,d) yields a topological space $(X, \mathscr{T})$ in which $\mathscr{T}$ consists of the open sets defined in terms of open balls $B(a,r)$. Each of the subsequent propositions corresponds to a *definition* in the realm of general topology. For example, a continuous mapping $f: X \to Y$ is one which has property (3) of Proposition 3. Likewise, if E is a non-empty subset of X, we take our lead from Proposition 4 and define a topology on E whose elements are all the sets $E \cap G$ for G in $\mathscr{T}$. Then E is a *compact* subset of X if, and only if, any covering of E by relatively open sets $E \cap G_i$ ($i \in I$) has a finite subcovering. The general Theorems 2, 5, 8 carry over without change to arbitrary topological spaces.

The mappings which preserve all the topological ideas are the *topological mappings* or *homeomorphisms*. We say that $f: X \to Y$ is a *homeomorphism* if f is a one–one continuous mapping which possesses a one–one continuous inverse mapping $g: Y \to X$. (For f, g to be inverses of each other we require

$$g(f(x)) = x,$$
$$f(g(y)) = y$$

for all x in X and all y in Y.)

We shall say no more about general topological spaces except to mention one property of 'separation' that holds in any metric space, viz.

O 4. *given two distinct points x, y in (X, d); there exist disjoint open sets A, B containing x, y, respectively.*

(See Ex. 19.) This property may or may not hold for a topological space – if it does, then the space is called a *Hausdorff* space, after the pioneer who first systematically used O 4 in conjunction with O 1, O 2, O 3 to define topological spaces [10]. In Exx. 20, 22 we have topological spaces which do not satisfy this property and hence are not *metrisable*, i.e. their open sets cannot be defined in terms of a metric.

Exercises

1. Show that A is closed if and only if $\overline{A} = A$.

2. (i) An *adherent* point of A is a point p such that every open set containing p contains at least one point of A. Show that the closure $\overline{A}$ consists of all the adherent points of A.

(ii) Show that $\overline{A}$ is obtained by adjoining to A all the limit points of A.

3. Show that the closure of $A \cup B$ is $\overline{A} \cup \overline{B}$.

4. Let the bounded interval I in $\mathbf{R}^k$ be defined by the inequalities $a_i \prec x_i \prec b_i$ ($i = 1, ..., k$) as in §4.1 and assume that I is not empty. Show that int I is defined by the inequalities $a_i < x_i < b_i$ and $\bar{I}$ is defined by the inequalities $a_i \leqslant x_i \leqslant b_i$.

5. Let (X,d) be a metric space. For any point x of X and any non-empty subset A of X, let
$$d(x,A) = \inf_{a \in A} d(x,a).$$
Use Ex. 2 (i) to show that $d(x,A) = 0$ if and only if $x \in \bar{A}$.

6. (i) In the notation of Ex. 5 show that
$$|d(x,A) - d(y,A)| \leqslant d(x,y)$$
for any x, y in X.
(ii) Let $f(x) = d(x,A)$ for any x in X; show that $f: X \to \mathbf{R}$ is continuous.

7. In the metric space (X,d) let $T = \{x \in X: d(x,a) \leqslant r\}$, $(a \in X, r > 0)$. Show that T is closed. What is the relationship of T to $B(a,r)$?

8. Show that every subset of a discrete metric space (Example 4) is both open and closed.

9. A *neighbourhood* of a point a of a metric (topological) space X is a subset of X which contains an open set containing a. Show that $f: X \to Y$ is continuous at the point a if and only if, for every neighbourhood N of $f(a)$, $f^{-1}(N)$ is a neighbourhood of a.

10. Let f, g be continuous real valued functions on a metric (topological) space X and c a real number. Show that $f+g, fg$ and cf are also continuous real valued functions on X.

11. Let $f: X \to Y$, $g: Y \to Z$ be continuous functions and define the composite function $h: X \to Z$ by the rule
$$h(x) = g(f(x)) \quad (x \in X).$$
Show that h is continuous.

12. Let $f: [a,b] \to \mathbf{R}$ be continuous. Show that the graph
$$\{(x, f(x)): x \in [a,b]\}$$
is a connected subset of $\mathbf{R}^2$. Hence prove the Intermediate Value Theorem by referring to Fig. A.9, and prove the Fixed Point Theorem for $[a,b]$ by referring to Fig. A.10.

13. Show that the mapping f given by $f(x) = 1/x$ $(0 < x < 1)$ is a homeomorphism of the bounded interval $(0,1)$ onto the unbounded interval $(1, \infty)$. This shows that boundedness in $\mathbf{R}$ is not a 'topological' property.

14. Show that a compact metric space which contains infinitely many points must have at least one limit point (Bolzano–Weierstrass).

15. Show that the union of a finite collection of compact sets is compact.

16. Let A be a symmetric $(k \times k)$ real matrix and define the quadratic form
$$f(x) = xAx^t \quad (x \in \mathbf{R}^k).$$
Show that f is a continuous function on $\mathbf{R}^k$. Let λ be the maximum value attained by f on the (compact) unit sphere $xx^t = 1$, at the point c. Show that
$$cA = \lambda c.$$
(This real number λ is the *maximum eigenvalue* of the matrix A.)

17. Let Y be a metric (topological) space. A *path* in Y is a continuous mapping $f: [a, b] \to Y$; $f(a)$ is the *initial point* and $f(b)$ is the *terminal point* of the path. We say that Y is *pathwise connected* if to each pair of points c, d of Y there is a path whose initial point is c and whose terminal point is d. Show that a pathwise connected space is connected. (There are examples of connected spaces that are not pathwise connected. See Mendelson [17].)

18. Show that a continuous mapping carries a pathwise connected space onto a pathwise connected space (cf. Theorem 2).

19. Show that a metric space satisfies the Hausdorff axiom O 4.

20. Find the four distinct topologies on the set $\{0, 1\}$. Which of these satisfy the Hausdorff axiom O 4?

21. Let X be a non-empty set and let $\mathscr{T}$ be the collection of all subsets of X. Show that this topology on X is the same as the one given by the discrete metric of Example 4.

22. Let $X = \{1, 2, 3, \ldots\}$. For each $n \geqslant 1$ let
$$O_n = \{n, n+1, n+2, \ldots\}.$$
Let $\mathscr{T} = \{\varnothing, O_1, O_2, \ldots\}$. Show that $(X, \mathscr{T})$ is a topological space. Does this space satisfy the Hausdorff axiom O 4?

23. Let X be a compact space and Y a Hausdorff space.
 (i) Show that any closed subset of X is compact (cf. the proof of Theorem 6, part (ii)).
 (ii) Show that any compact subset of Y is closed (cf. the proof of Theorem 7, part (ii)).
 (iii) If f is a one-one continuous mapping of X onto Y show that f is a homeomorphism.

SOLUTIONS

§1.1

1. Suppose that $n \leqslant K$ for all integers n. According to the Axiom of Completeness the increasing sequence $\{n\}$ must then converge to a real limit, s, say. Take $\epsilon = 1$ in the definition of convergence; there exists an integer N such that $s-1 < n < s+1$ for $n \geqslant N$. In particular $s-1 < N$ and so $s+1 < N+2$ – a contradiction!

2. Given $\epsilon > 0$; there exists an integer N such that
$$s-\epsilon < s_n < s+\epsilon, \quad t-\epsilon < t_n < t+\epsilon$$
for $n \geqslant N$. From this it follows that
$$\max\{s,t\}-\epsilon < \max\{s_n,t_n\} < \max\{s,t\}+\epsilon$$
for $n \geqslant N$. Similarly for min.

3. As s_n is also of the form (integer/10^n) it is clear that $s_n \leqslant s_{n+1}$. Let $t_n = s_n + 1/10^{n-1}$. Then t_n is the smallest (integer/10^{n-1}) whose square $\geqslant 2$, and so $\{t_n\}$ is decreasing. By the Axiom of Completeness there is a real number s such that $s_n \to s$ and $t_n \to s$. Now $s_n^2 < 2$ and $t_n^2 \geqslant 2$ yield $s^2 \leqslant 2$ and $s^2 \geqslant 2$, whence $s^2 = 2$.

4. Let $a_0.a_1a_2\ldots = b_0.b_1b_2\ldots$ and let i be the smallest integer for which $a_i \neq b_i$. By subtracting $a_0.a_1a_2\ldots a_{i-1}$ from each side and multiplying by 10^i we may arrange that $i = 0$. Without loss of generality we may also assume that $a_0 > b_0$ and so $a_0 \geqslant b_0 + 1$. Now
$$a_0.a_1a_2\ldots \geqslant a_0 \geqslant b_0+1 \geqslant b_0.b_1b_2\ldots.$$
But the extremes are equal and so
$$a_0.a_1a_2\ldots = a_0, \quad b_0.b_1b_2\ldots = b_0+1.$$
This can only happen if $a_1 = a_2 = \ldots = 0$ and $b_1 = b_2 = \ldots = 9$.

5. By the Axiom of Archimedes we may find an integer N greater than k (if $k < 0$ then $N = 0$ will do). By the same argument there are integers $n \geqslant N-k$: all these integers n are positive and so, by the Well-Ordering Axiom, there is a smallest, n_0, say. It is now easy to verify that $N-n_0$ is the largest integer $\leqslant k$.

§1.2

2. If $c \in X \cup Y$ then either $c \leqslant \sup X$ or $c \leqslant \sup Y$. Hence
$$\sup(X \cup Y) \leqslant \max\{\sup X, \sup Y\}.$$
On the other hand it is clear that
$$\sup X \leqslant \sup(X \cup Y) \quad \text{and} \quad \sup Y \leqslant \sup(X \cup Y),$$
whence
$$\max\{\sup X, \sup Y\} \leqslant \sup(X \cup Y).$$

For any x in X, y in Y, $x+y \leq \sup X + \sup Y$, and so
$$\sup Z \leq \sup X + \sup Y.$$
For any $\epsilon > 0$, we may find x in X, y in Y such that
$$x > \sup X - \epsilon, \quad y > \sup Y - \epsilon;$$
thus $\quad\sup Z > \sup X + \sup Y - 2\epsilon.$
As ϵ is arbitrary, it follows that
$$\sup Z \geq \sup X + \sup Y.$$

3. Find n_0, n_1 as in Ex. 1.1.5. Now S is a non-empty set of positive integers and so contains a smallest member $n_0 - n_2$, say. Then n_2 is the greatest integer lower bound for X.

§ 1.3

4.
$$a(0+0) = a0 + a0, \tag{9}$$
$$\Rightarrow \quad a0 = a0 + a0, \tag{3}$$
$$\Rightarrow \quad a0 + (-a0) = (a0 + a0) + (-a0), \tag{4}$$
$$\Rightarrow \quad a0 + (-a0) = a0 + (a0 + (-a0)), \tag{2}$$
$$\Rightarrow \quad 0 = a0 + 0, \tag{4}$$
$$\Rightarrow \quad 0 = a0. \tag{3}$$

If $ab = 0$ and $a \neq 0$ then $a^{-1}(ab) = 0$ by the first part. Thus
$$(a^{-1}a)b = 0, \tag{6}$$
$$\Rightarrow \quad 1b = 0, \tag{5), (8}$$
$$\Rightarrow \quad b = 0. \tag{5), (7}$$

8. By (12) $a \geq c$. If we had $a = c$ then we should have $a > b$ and $b > a$, whence $a = b$ by (11) – contrary to $a > b$.

9. In any field $a^2 = (-a)^2$ (Ex. 6), and so in any ordered field $a^2 \geq 0$ by (14).
Suppose that $\mathbf{F}_p$ is ordered; then $1 > 0$ ($1 = 1^2$) and $1+1 > 1$. By Ex. 8, $1+1 > 0$, $(1+1)+1 > 0$, ... This yields a contradiction at the p-th step. In $\mathbf{C}, i^2 = -1 < 0$.

11. The definition of convergence is as before (p. 3). Let $\{s_n\}$ have least upper bound s. For any $\epsilon > 0$, $s - \epsilon < s$ is too small to be an upper bound and so there is an integer N such that $s_N > s - \epsilon$. As $\{s_n\}$ is increasing and bounded above by s,
$$s - \epsilon < s_n \leq s$$
for all $n \geq N$. Hence $s_n \to s$.

SOLUTIONS

12. If $X = \{1, 2, 3, \ldots\}$ were bounded then X would have a least upper bound s, say. As $s-1 < s$ there would be a positive integer N such that $s-1 < N$; but then $s < N+1$ would contradict the fact that s is an upper bound of X.

13. The set A is not empty and is bounded above by each element of B (also not empty). Let $c = \sup A$ and check that $a \leqslant c \leqslant b$ for all a in A, b in B.

14. Let $I_n = [a_n, b_n]$; then $a_n \leqslant b_1$ for all n. Let $c = \sup\{a_n\}$ and verify that $a_n \leqslant c \leqslant b_n$ for all n.

If $l(I_n) \to 0$ then $\{a_n\}, \{b_n\}$ both converge to c. (We assume that a sequence cannot converge to more than one limit; this can be proved by means of Proposition 1.1.1 (v).)

§ 2.1

1. The subset S of $\mathbf{N}$ is defined as follows:
$$n \in S \quad \text{if and only if} \quad n \notin S_n.$$

§ 2.2

1. For given $\epsilon > 0$, suppose that the set S is covered by the sequence $\{I_n\}$ of bounded intervals of total length less than ϵ. Replace each interval I_n by an open interval J_n with the same centre and of length $l(I_n) + \epsilon/2^n$. The total length of the sequence $\{J_n\}$ is less than 2ϵ.

§ 3.1

2. Using Proposition 1 let $\phi = a_1 \chi_{I_1} + \ldots + a_r \chi_{I_r}$ where $I_1, \ldots, I_r$ are disjoint. Then $|\phi| = |a_1|\chi_{I_1} + \ldots + |a_r|\chi_{I_r}$.

3. Use Ex. 2 and the identities
$$2 \max\{f, g\} = f + g + |f-g|,$$
$$2 \min\{f, g\} = f + g - |f-g|.$$

4. Use Ex. 2 and the identities $2f^+ = |f| + f$, $2f^- = |f| - f$.

5.
$$\chi_{A \cup B} = \max\{\chi_A, \chi_B\}, \quad \chi_{A \cap B} = \min\{\chi_A, \chi_B\},$$
$$\chi_{A \setminus B} = (\chi_A - \chi_B)^+, \quad \chi_{A \triangle B} = |\chi_A - \chi_B|.$$

6. Using Proposition 1 let the step function $\phi = a_1 \chi_{I_1} + \ldots + a_r \chi_{I_r}$, where $I_1, \ldots, I_r$ are disjoint. Then $\phi = \chi_S$ for some set S if and only if each a_i is 0 or 1, and this is true if and only if S is in $\mathscr{R}$.

If χ_S, χ_T are step functions we may use Ex. 5 to see that $\chi_{S \cup T}$, etc. are step functions.

By Proposition 3, $\chi_S \leqslant \chi_T \Rightarrow l(S) \leqslant l(T)$.

7. Using Proposition 1 let $\phi = a_1\chi_{I_1} + \ldots + a_r\chi_{I_r}$, where $I_1, \ldots, I_r$ are disjoint. Then S is the union of those I_i for which $a_i \geq k$ (if there are none, then S is empty). As $\phi \geq 0$,
$$\chi_S \leq \frac{1}{k}\phi$$
and so by Proposition 3 $\quad l(S) \leq \dfrac{1}{k}\displaystyle\int \phi.$

8. According to Ex. 7 each $S_n \in \mathscr{R}$. As $\phi_n \leq \phi_{n+1}$ it follows that $S_n \subset S_{n+1}$. By Ex. 7
$$l(S_n) \leq \frac{1}{k}\int \phi_n;$$
thus $\{l(S_n)\}$ is increasing (Ex. 6) and bounded above, hence convergent by the Axiom of Completeness.

9. As in the proof of Proposition 2 find disjoint intervals $K_1, \ldots, K_t$ so that
$$\phi = a_1\chi_{K_1} + \ldots + a_t\chi_{K_t}, \quad \psi = b_1\chi_{K_1} + \ldots + b_t\chi_{K_t}.$$
Then $\quad \phi\psi = a_1 b_1\chi_{K_1} + \ldots + a_t b_t\chi_{K_t}.$

§ 3.2

1. For the second integral, divide $[0, 1)$ into $r = 2^n$ equal subintervals I_i of type $[\,,\,)$ and let
$$c_i = \inf_{x \in I_i} x^2.$$
Then the step function $\quad \phi_n = c_1\chi_{I_1} + \ldots + c_r\chi_{I_r}$

approximates the integrand from below (see the first few paragraphs of § 3.3 and Fig. S. 1)

$$\int \phi_n = \{0 + 1^2/r^2 + 2^2/r^2 + \ldots + (r-1)^2/r^2\}/r = (r-1)\,r(2r-1)/6r^3$$

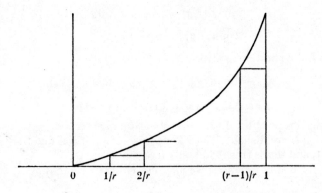

Fig. S. 1

using a standard formula for the sum of squares. Successive subdivision gives an increasing sequence $\{\phi_n\}$ of step functions which converges to the integrand everywhere, and $\int \phi_n \to \frac{1}{3}$ as n (and hence r) tends to ∞.

2. As $\chi_S = \chi_{[0,1]}$ a.e., $\int \chi_S = 1$.

5. Let $f = g - h$ where $g, h \in L^{\text{inc}}$. Then there exist increasing sequences $\{\psi_n\}, \{\theta_n\}$ such that $\psi_n \to g$, $\theta_n \to h$ almost everywhere and

$$\int \psi_n \to \int g, \quad \int \theta_n \to \int h.$$

Take $\phi_n = \psi_n - \theta_n$; then $\phi_n \to f$ almost everywhere and

$$\int |\phi_n - f| \leq \int |\psi_n - g| + \int |h - \theta_n| \to 0.$$

6. First of all, let $f \in L^{\text{inc}}$ and $\phi_n \uparrow f$ a.e. with bounded $\int \phi_n$. Then

$$\phi_n \chi_I \uparrow f \chi_I \text{ a.e.}$$

with bounded $\int \phi_n \chi_I$ and so $f \chi_I \in L^{\text{inc}}$. This extends by linearity to the case where $f \in L^1$ and then to the case where χ_I is replaced by a step function.

7. If $f \chi_J \in L^1(\mathbf{R})$ then $f \chi_I \in L^1(\mathbf{R})$ by Ex. 6.

10. If $k > 0$ ($k < 0$) the graph of ψ is obtained by translating the graph of ϕ a distance $|k|$ to the left (right), whence $\int \psi = \int \phi$.

The graph of θ is obtained from the graph of ϕ by a 'horizontal change of scale' in which the distance of each point from the y-axis is multiplied by $1/|k|$; if $k < 0$ there is also a reflection in the y-axis. Thus

$$\int \theta = \frac{1}{|k|} \int \phi.$$

(This may be stated more formally in terms of intervals.)

The result for f in L^{inc} follows by taking limits and then extends at once to L^1 by linearity.

If $f \in L^1[a,b]$ then
$$\int_{a-k}^{b-k} f(x+k)\,dx = \int_a^b f(x)\,dx,$$

$$\int_{a/k}^{b/k} f(kx)\,dx = \frac{1}{k} \int_a^b f(x)\,dx$$

(due account being taken of the convention in Ex. 9).

11. Let $S_n = I_1 \cup \ldots \cup I_n$. Then $\{\chi_{S_n}\}$ is an increasing sequence of step functions which converges to χ_S and $\int \chi_{S_n} \leq \frac{1}{2}$. Thus $\chi_S \in L^{\text{inc}}$ and $\int \chi_S \leq \frac{1}{2}$. Hence $f \in L^1$ and $\int f = 1 - \int \chi_S \geq \frac{1}{2}$.

For each rational point p in $(0, 1)$ there is an open interval containing p on which f is zero; any interval I with $l(I) > 0$ therefore contains a non-empty open interval on which f is zero. In view of this, any step function ϕ_n which satisfies $\phi_n \leq f$ a.e. must also satisfy $\phi_n \leq 0$ a.e. (even $\phi_n \leq 0$ outside a finite set of points). If f were to belong to L^{inc} this would imply that $\int f \leq 0$ contrary to the fact that $\int f \geq \frac{1}{2}$.

§ 3.3

1. The proof is very similar to that of Proposition 1.1.1.

2. We only need to verify that the polynomial functions f defined by $f(x) = c$ and $f(x) = x$ are continuous and apply Ex. 1.

3. According to Ex. 2 the integrand is continuous on the appropriate interval of integration $[1, x]$ or $[x, 1]$ and we may apply Corollary 1.

If $x \geq 1$ then $1/x \leq 1/t \leq 1$ for $1 \leq t \leq x$ and so we may apply Proposition 3.2.2. The proof for $0 < x < 1$ is similar but uses the convention of Ex. 3.2.9.

4. As in Ex. 3, $$\frac{1}{r} \leq \int_{r-1}^{r} \frac{1}{t} dt \leq \frac{1}{r-1}.$$

To verify that these inequalities are *strict* we may introduce a further point of subdivision as in Fig. S. 2: the shaded portion has strictly positive area. Then sum from $r = 2$ to $r = n$.

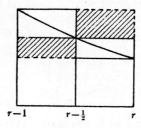

Fig. S. 2

5. Given any $\epsilon > 0$; there is an $\eta > 0$ such that
$$|f(y) - f(g(p))| < \epsilon \quad \text{for all } y \text{ satisfying} \quad |y - g(p)| < \eta.$$
Then there is a $\delta > 0$ such that
$$|g(x) - g(p)| < \eta \quad \text{for all } x \text{ satisfying} \quad |x - p| < \delta.$$
Combine these by setting $y = g(x)$: we have
$$|h(x) - h(p)| < \epsilon \quad \text{for all } x \text{ satisfying} \quad |x - p| < \delta.$$

The Riemann Integral

A. (i) As $\int\phi \leqslant \int\psi$ for any step functions ϕ, ψ satisfying $\phi \leqslant f \leqslant \psi$ and ϕ, ψ may be varied independently, we deduce that

$$\sup\int\phi \leqslant \inf\int\psi.$$

(ii) Take $\phi = f = \psi$.

(iii) $\inf\int(\psi - \phi) = \inf\int\psi - \sup\int\phi$ (cf. Ex. 1.2.2).

B. $\{\int\phi_n\}$ is increasing and bounded above by $\int\psi_1$; thus $\int\phi_n \to l$, say. But $\int\psi_n - \int\phi_n \to 0$ and so $\int\psi_n \to l$ by Proposition 1.1.1. Now

$$\int\phi_n \leqslant \sup\int\phi \leqslant \inf\int\psi \leqslant \int\psi_n$$

for all n and so

$$\sup\int\phi = \inf\int\psi = l \quad \text{by the 'squeeze'}.$$

C. (i) By the definition of the lower Riemann integral as a supremum, we may find a step function $\theta \leqslant f$ vanishing outside $[a, b]$ such that

$$\int\theta > \underline{\int_a^b} f(x)\,dx - \tfrac{1}{2}\epsilon.$$

Express θ in terms of disjoint subintervals of $[a, b]$ and let P be the corresponding partition of $[a, b]$. The lower approximating step function ϕ determined by P (as above) satisfies $\theta \leqslant \phi \leqslant f$.

(ii) Let J be an interval of the partition P_n. If J is contained in one of the intervals $I_1, \ldots, I_r$, then

$$\int_J \phi_n \geqslant \int_J \phi.$$

On the other hand if J contains points of more than one of the I's then, at the very worst,

$$\int_J \phi_n \geqslant \int_J \phi - 2Kl(J).$$

This overlapping occurs at most $(r-1)$ times and $l(J) \leqslant \nu_n$, so the given inequality holds.

(iii) Choose ϕ as in (i); then choose N (depending on ϕ and ultimately on ϵ) such that

$$(r-1)\,2K\nu_n < \tfrac{1}{2}\epsilon \quad \text{whenever} \quad n \geqslant N,$$

and combine the inequalities of (i) and (ii).

This now establishes Darboux's Theorem for the lower integrals and the upper integrals are treated in virtually the same way.

D. As P_{n+1} refines P_n for $n \geqslant 1$ it follows that $\{\phi_n\}$ is increasing, $\{\psi_n\}$ is decreasing and $\phi_n \leqslant f \leqslant \psi_n$ for all n. Finally, by Darboux's Theorem, $\int(\psi_n - \phi_n) \to 0$.

E. (i) $\inf\{f(x): x \in I_n\}$ increases with n and is bounded above by K, hence converges to $g(p)$, say.

Let $\{J_n\}$ be another decreasing sequence of bounded intervals each containing p as an interior point such that $l(J_n) \to 0$, and let $g_1(p)$ be the corresponding limit. For any $n \geq 1$, p is an interior point of I_n and has distance $\delta_n > 0$ from the nearest end point of I_n. Let m be large enough to ensure that $l(J_m) < \delta_n$; then $J_m \subset I_n$ and so

$$\inf\{f(x): x \in I_n\} \leq \inf\{f(x): x \in J_m\} \leq g_1(p).$$

Hence $g(p) \leq g_1(p)$. The reverse inequality follows by interchanging the roles of the I's and J's.

The argument for $h(p)$ is similar.

(ii) Immediate from the definitions.

(iii) If f is continuous at p, follow the proof of Theorem 1 to show that $g(p) = h(p)$.

If $g(p) = h(p)$ then each equals $f(p)$. To any $\epsilon > 0$, we may find N such that

$$\inf\{f(x): x \in I_N\} > f(p) - \epsilon$$

$$\sup\{f(x): x \in I_N\} < f(p) + \epsilon.$$

As p is an interior point of I_N we may find $\delta > 0$ such that $(p-\delta, p+\delta) \subset I_N$ and so

$$f(p) - \epsilon < f(x) < f(p) + \epsilon \quad \text{for all } x \text{ in} \quad (p-\delta, p+\delta).$$

In other words f is continuous at p.

(iv) In each case $h = \chi_{(a,b)}$. In the first two cases $g = \chi_{(a,b)}$ and in the third, $g = 0$.

F. (i) Direct verification.

(ii) The argument given for Ex. E(i) shows that $g(p) \leq \lim \phi_n(p)$ with equality if p is an interior point of all the intervals J_n.

(iii) Direct verification.

(iv) Using Theorem 5.1.1, Corollary we deduce that $h - g = 0$ a.e. and so $f = g = h$ a.e. Finally $\int f = l$ by Theorem 3.2.3.

§ 3.4

3. $$\int_1^{xy} \frac{1}{t} dt = \int_1^{x} \frac{1}{t} dt + \int_x^{xy} \frac{1}{t} dt \quad \text{(Ex. 3.2.9)}.$$

In the last integral substitute $t = xu$ (x is regarded as a constant): by Proposition 2 it equals

$$\int_1^y \frac{1}{xu} x\, du = \int_1^y \frac{1}{u} du.$$

4. Integrating by parts

$$\int_1^2 \log x \, dx = [x \log x]_1^2 - \int_1^2 dx = 2\log 2 - 1.$$

SOLUTIONS

5. Summing for $i = 1, \ldots, n$ and using Ex. 3

$$\log\left(\frac{a_1 \ldots a_n}{A^n}\right) \leq \frac{nA}{A} - n$$

or

$$n \log \frac{G}{A} \leq 0.$$

7. The proof of Proposition 2 applies verbatim except that F_0 is defined for x in I. The proviso about the interval I ensures that $f(G(t))$ is *defined* for all t in $[c, d]$.

Substitute $x = \sin \theta$. The given integral equals

$$\int_0^0 x^2 \, dx = 0.$$

(Apparently trivial, but note that the interval I in this case is $[0, 1]$!)

8. The sign of the square root must be specified carefully.

§ 3.5

1. $$|F(x_i) - F(x_{i-1})| = \left|\int_{x_{i-1}}^{x_i} f\right| \leq \int_{x_{i-1}}^{x_i} |f|.$$
Sum for $i = 1, \ldots, r$.

2. $$\Sigma|F(x_i) - F(x_{i-1})| \leq \Sigma\{G(x_i) - G(x_{i-1})\} + \Sigma\{H(x_i) - H(x_{i-1})\}$$
$$= G(b) - G(a) + H(b) - H(a).$$

3. $$p - n = \sum_{i=1}^{r} (F(x_i) - F(x_{i-1})) = F(b) - F(a),$$

$$p + n = \sum_{i=1}^{r} |F(x_i) - F(x_{i-1})|.$$

Rewrite the first as $p = n + (F(b) - F(a))$ and use Ex. 1.2.2.

4. As in Ex. 3, $P(x) - N(x) = F(x) - F(a)$ and P, N are increasing on $[a, b]$.

5. Use condition (6) with $\tfrac{1}{2}\epsilon$ in place of ϵ and consider the positive and negative terms $F(b_i) - F(a_i)$ separately.

6. Both follow immediately from the inequality

$$\Sigma|F(x_i) - F(x_{i-1})| \leq K\Sigma(x_i - x_{i-1}).$$

8. To check the continuity of F use only one interval (a_i, b_i) in (6). Take $\epsilon = 1$ in the condition of Ex. 5 and find a corresponding $\delta > 0$. Let N be the smallest integer greater than $(b-a)/\delta$ and subdivide $[a, b]$ into N equal intervals. Given any subdivision $a = x_0 < x_1 < \ldots < x_r = b$, we may combine these two subdivisions and so split $[a, b]$ into N sets of intervals,

each set being of total length less than δ. Extra points of division can only increase the sum and so
$$\Sigma |F(x_i) - F(x_{i-1})| < N.$$

10. The given rules define F at the points of trisection $m/3^n$ of $(0,1)$; moreover, if p, q are two such points, $p \leq q \Rightarrow F(p) \leq F(q)$. Let $F(0) = 0$, $F(1) = 1$. Any point x of $[0,1]$ has *ternary* expansion $x = 0 \cdot a_1 a_2 \ldots$ (prefer a finite expansion to one which ends in a string of 2's: cf. Ex. 1.1.4). The partial sums $x_n = 0 \cdot a_1 a_2 \ldots a_n$ are points of trisection and we may set $F(x) = \lim F(x_n)$. This definition preserves the order, i.e. F is an increasing function on $[0,1]$. According to Proposition 3.3.1 the only discontinuities of F are simple jumps; but the repeated bisection of the vertical interval does not allow any room for jumps, and so F is continuous.

Let S_n be the set of 2^n disjoint closed intervals each of length $1/3^n$ described in §2.3, and Cantor's ternary set $S = \bigcap S_n$. If $p \in [0,1] \backslash S$, then $p \in [0,1] \backslash S_n$ for some n; this means that F is constant on some open interval containing p and so $f(p) = F'(p) = 0$. As $l(S_n) = (\tfrac{2}{3})^n \to 0$, S is null and $f = 0$ a.e. Thus
$$\int_0^1 f(x)\, dx = 0.$$

Denote the intervals of S_n by $[a_i, b_i]$ $i = 1, 2, \ldots, 2^n$. As F is constant on each interval of $[0,1] \backslash S_n$ it follows that
$$\Sigma\{F(b_i) - F(a_i)\} = 1 \quad \text{although} \quad \Sigma(b_i - a_i) = (\tfrac{2}{3})^n.$$

This rules out the possibility that F be absolutely continuous.

§4.1

6. For $c < a < b < d$,
$$[a, b) = (c, b) \backslash (c, a),$$
$$(a, b] = (a, d) \backslash (b, d),$$
$$[a, b] = \{(c, d) \backslash (c, a)\} \backslash (b, d).$$

Express these relations in terms of characteristic functions and use the first part.

§4.2

1. Let I_n be the closed interval defined by $|x_1| \leq n, \ldots, |x_k| \leq n$ and H the given hyperplane. We have already seen that $I_n \cap H$ is null for each $n \geq 1$. By Proposition 2 the union $\bigcup (I_n \cap H)$ is null, i.e. H is null.

2. For a given rational x the line consisting of all (x, y) $(y \in \mathbf{R})$ is null (Ex. 1); similarly with x, y interchanged. Then use Proposition 2.

3. Using Ex. 2, $\int \chi_S = 1$.

6, 7. See the proof of Proposition 6.3.1.

SOLUTIONS

8. Consider a regular polygon with n sides inscribed in the given circle and let s be the length of each side. At each vertex V the closed square, centre V, side $2s$ has area $4s^2$. There are n such squares which cover the circle and have total area $A_n = 4ns^2$. But $s < 2\pi r/n$ and so $A_n < 16\pi^2 r^2/n$. Now let $n \to \infty$.

The characteristic functions χ_S, χ_T are continuous except at the points of the circle $|x| = r$. Thus by Theorem 4, χ_S, χ_T are integrable, and by Theorem 3 their integrals are equal.

9. See the proof of Proposition 6.3.1.

§ 4.3

1. Let I, J be bounded intervals in $\mathbf{R}^l, \mathbf{R}^m$, respectively, and $K = I \times J$ the bounded interval in $\mathbf{R}^k$ consisting of all points (x, y) with x in I, y in J. Then
$$\chi_K(x, y) = \chi_I(x)\chi_J(y)$$
and
$$\int \chi_K(x, y)\, dy = m(J)\chi_I(x).$$

Any step function on $\mathbf{R}^k$ is of the form
$$\phi = c_1 \chi_{K_1} + \ldots + c_t \chi_{K_t}.$$
In the above notation
$$\Phi = c_1 m(J_1) \chi_{I_1} + \ldots + c_t m(J_t) \chi_{I_t}.$$

2. (i) The cross-section at height z $(0 \leq z \leq 2)$ is a disk whose area is $\pi(2-z)$.
$$\text{Volume} = \int_0^2 \pi(2-z)\, dz = 2\pi.$$
(ii) The cross-section at height z has area $\tfrac{1}{2}(1-z)^2$.
$$\text{Volume} = \int_0^1 \tfrac{1}{2}(1-z)^2\, dz = \tfrac{1}{6}.$$

3. Substitute $y = x \tan\theta$. Then
$$\int_0^1 \frac{x^2 - y^2}{(x^2+y^2)^2}\, dy = \frac{1}{x}\int_0^{\tan^{-1} 1/x} \frac{(1 - \tan^2\theta)}{(1 + \tan^2\theta)^2} \sec^2\theta\, d\theta$$
$$= \frac{1}{x}\int_0^{\tan^{-1} 1/x} (\cos^2\theta - \sin^2\theta)\, d\theta$$
$$= \frac{1}{x}\left[\frac{\sin 2\theta}{2}\right]_0^{\tan^{-1} 1/x} = \frac{1}{1+x^2}.$$
$$\int_0^1 \frac{dx}{1+x^2} = [\tan^{-1} x]_0^1 = \frac{\pi}{4}.$$

Interchanging x, y alters the sign of f and so the other repeated integral is $-\pi/4$.

In view of the Corollary to Fubini's Theorem, $f \notin L^1(I)$.

5. (i) See Exx. 4.2.1, 8.

(ii) In view of (i) we may as well assume that the triangle S is open (i.e. has the sides removed). To be sure that S has an area we may approximate S by an increasing sequence of 'escalators' (Figs. S. 3, 4):

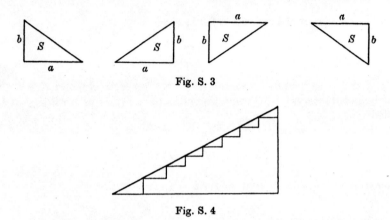

Fig. S. 3

Fig. S. 4

or simply quote the general result of Proposition 6.3.1, already mentioned in the text. Interpreting the area as the integral of cross-sections, as on p. 88, we have the area

$$m(S) = \int_0^a \frac{b}{a} x\, dx = \tfrac{1}{2}ab.$$

(iii)

Fig. S. 5

The area of the given rectangle (Fig. S. 5) is

$(a \cos \theta + b \sin \theta)(a \sin \theta + b \cos \theta) - a^2 \cos \theta \sin \theta - b^2 \cos \theta \sin \theta = ab.$

(iv) The area (Lebesgue measure) of a set S in $\mathbf{R}^2$ is defined as the integral $\int \chi_S$. This is defined (using sequences of step functions) in terms of the areas of bounded intervals; but (iii) shows that these areas are unaltered by translations, rotations and reflections, and so the same is true for all sets S of (finite) Lebesgue measure.

§ 5.1

1. Let S be the null set consisting of all points x for which $\{f_n(x)\}$ is not increasing. Apply Theorem 1 to the sequence $\{f_n(1-\chi_S)\}$.

2. If S is null, then $\chi_S = 0$ a.e. and $m(S) = 0$. If $m(S) = 0$ the Corollary to the Monotone Convergence Theorem shows that $\chi_S = 0$ a.e., i.e. S is null.

3. Let $f = g_1 - h_1$ where $g_1, h_1 \in L^{\text{inc}}$; then there exist increasing sequences $\{\phi_n\}, \{\psi_n\}$ of step functions converging a.e. to g_1, h_1. Let $\theta_1 = \min\{\phi_1, \psi_1\}$ and $g = g_1 - \theta_1$, $h = h_1 - \theta_1$ so that $f = g - h$ where $g, h \geq 0$ a.e. Then $f = g^+ - h^+$ a.e. and $g^+, h^+ \in L^{\text{inc}}$.

4. In each case take $I_n = [0, n]$.
$$\int_0^n e^{-x}dx = 1 - e^{-n} < 1 \quad \text{and so} \quad \int_0^\infty e^{-x}dx = 1$$
by Proposition 1. For $n \geq 2$
$$\int_0^n e^{-x^2}dx \leq \int_0^1 e^{-x^2}dx + \int_1^n e^{-x}dx < 1 + e^{-1}.$$
This establishes the existence of the second integral.

In $\mathbf{R}^2$ let $J_n = I_n \times I_n$ (cf. the solution for Ex. 4.3.1). As J_n is bounded and closed
$$\int_{J_n} e^{-(x^2+y^2)}d(x,y)$$
exists, and by Fubini's Theorem equals
$$\left(\int_{I_n} e^{-x^2}dx\right)\left(\int_{I_n} e^{-y^2}dy\right).$$
Let $n \to \infty$ and apply Proposition 1.

[In Ex. 6.4.4 we use the polar transformation to show that $\alpha^2 = \frac{1}{4}\pi$.]

6. (i) $$\int_{1/n}^1 e^{-x}x^{\alpha-1}dx \leq \int_{1/n}^1 x^{\alpha-1}dx < 1/\alpha.$$
Apply Proposition 1 with $I_n = (1/n, 1)$.

The expansion of e^x as a power series in x contains higher powers than $x^{\alpha+1}$. Thus we may find $K > 1$ such that $e^{-x}x^{\alpha-1} < x^{-2}$ for all $x > K$. It follows that
$$\int_1^n e^{-x}x^{\alpha-1}dx \leq \int_1^K e^{-x}x^{\alpha-1}dx + \int_K^n x^{-2}dx$$
for $n \geq K$ and we may apply Proposition 1 with $I_n = (1, n)$.

(ii) Consider the $n+1$ (strictly positive) numbers

$$1-\frac{x}{n}, \quad 1-\frac{x}{n}, \ldots, \quad 1-\frac{x}{n}, \quad 1.$$

(iii) From the inequalities of Ex. 3.3.3

$$\frac{-x}{1-x/n} \leq n\log\left(1-\frac{x}{n}\right) \leq -x.$$

Thus $\log\left(1-\frac{x}{n}\right)^n \to -x$ as $n \to \infty$. By the continuity of the exponential (inverse of the logarithm)

$$\left(1-\frac{x}{n}\right)^n \to e^{-x} \quad \text{as} \quad n \to \infty.$$

In view of (i) and (ii) the Monotone Convergence Theorem gives the required limit.

7. By the Axiom of Completeness $\{f_n\}$ converges (everywhere) to a positive limit function f. By the Monotone Convergence Theorem $f \in L^1$ and $\int f = 0$; thus $f = 0$ a.e.

8. The series $\Sigma \int a_n^+, \Sigma \int a_n^-$ are convergent by comparison with the given convergent series. By the Monotone Convergence Theorem Σa_n^+, Σa_n^- are convergent a.e. to functions g, h in L^1. As $a_n = a_n^+ - a_n^-$ and $|a_n| = a_n^+ + a_n^-$, Σa_n is absolutely convergent a.e. to the function $f = g - h$ in L^1.

9. For $k, n \geq 1$,

$$\int_0^k e^{-nx} x^{\alpha-1} dx = \int_0^{nk} e^{-t} t^{\alpha-1} dt / n^\alpha \quad (\text{substitute } t = nx).$$

Thus
$$\int_0^\infty e^{-nx} x^{\alpha-1} dx = \Gamma(\alpha)/n^\alpha \quad \text{for} \quad \alpha > 0.$$

Now $\sum_{n=1}^{\infty} e^{-nx}$ converges to the sum $e^{-x}/(1-e^{-x})$ for all $x > 0$, and $\sum_{n=1}^{\infty} 1/n^\alpha$ converges for $\alpha > 1$. The Monotone Convergence Theorem therefore gives the required equation.

§ 5.2

1. Taking $m = N$, $s_N - \epsilon < s_n < s_N + \epsilon$ for all $n \geq N$. The argument of Lemma 1 shows that $\{s_n\}$ is bounded. Moreover

$$s_N - \epsilon \leq l_N \leq u_N \leq s_N + \epsilon$$

from which
$$s_N - \epsilon \leq l \leq u \leq s_N - \epsilon$$

and so
$$0 \leq u - l \leq 2\epsilon.$$

SOLUTIONS 255

As ϵ is arbitrarily small this implies that $l = u$ and we may apply Proposition 1.

2. $f_n(x) \leq \frac{1}{2}$ for $0 \leq x \leq 1$. By considering the intervals $(0, 1/n]$, $(1/n, 1]$ separately we see that $g_n(x) < x^{-\frac{1}{2}}$, and so $h_n(x) < x^{-\frac{3}{4}}$, for $0 < x \leq 1$.

4. $f_n = \chi_{(0, n]}$.

5. In the notation used in the proof of Theorem 2,

$$l_{nk} \leq f_n, \ldots, l_{nk} \leq f_{n+k}.$$

Thus $$0 \leq \int l_{nk} \leq \min\left\{\int f_n, \ldots, \int f_{n+k}\right\}.$$

As $k \to \infty$ this gives

$$0 \leq \int l_n \leq \inf\left\{\int f_n, \int f_{n+1}, \ldots\right\}$$

and as $n \to \infty$ this in turn gives

$$0 \leq \int f \leq \liminf \int f_n.$$

Consider the bounded interval $[0, 1]$. Let $f_n(x) = n(1 - nx)$ for $0 \leq x \leq 1/n$, and $f_n(x) = 0$ otherwise. For each $n \geq 1$ the area under the graph is equal to $1/2$, i.e. $\int f_n = \frac{1}{2}$, but $f_n(x) \to 0$ for $0 < x \leq 1$.

A 'smoother' example with the same idea would be given by

$$f_n(x) = nxe^{-nx^2} \quad \text{for} \quad 0 \leq x \leq 1.$$

6. Integrate by parts. As $I_{0,\alpha+n} = 1/(\alpha+n)$, the reduction formula gives

$$I_{n,\alpha} = \frac{n!}{\alpha(\alpha+1)\ldots(\alpha+n)}.$$

7. $F'(x) = f(x) = 2x \sin 1/x^2 - (2/x) \cos 1/x^2$ for $x > 0$. The first term is dominated by x and so is integrable on $(0, 1)$. On the other hand

$$\int_{1/n}^{1} (2/x) \cos 1/x^2 \, dx = \int_{1}^{n^2} \frac{\cos v}{v} \, dv$$

(substitute $v = 1/x^2$) and we may apply the argument of p. 100 to show that

$$\int_{1}^{\infty} \frac{|\cos v|}{v} \, dv$$

does not exist as a Lebesgue integral.

8. Simple verification of special cases.

9. Let $|f| \leq K$. By Ex. 3.2.5, f is the limit a.e. of a sequence $\{\phi_n\}$ of step functions. By Ex. 3.2.6, $\{\phi_n g\}$ is a sequence of integrable functions which converges a.e. to fg. As $|fg| \leq K|g|$ we may apply Proposition 2.

For example, let $f = g$ where $f(x) = x^{-\frac{1}{2}}$ for $0 < x < 1$, and $f(x) = 0$ otherwise.

10. $f_n(x) \to f(x)$ for $a \leq x < b$ (and $f_n(b) = 0$). Let $x \in [a, b)$; if $x + 1/n \leq b$, then $f_n(x) = f(c)$ for some c in (a, b) and so $|f_n(x)| \leq K$. Even if $x + 1/n > b$ we still have $|f_n(x)| = |n(F(b) - F(x))| \leq |f(c)|$ for some c in (a, b) and
$$|f_n(x)| \leq K.$$
The given limit therefore follows by the Bounded Convergence Theorem.

By a simple translation,
$$\int_a^b F\left(x + \frac{1}{n}\right) dx = \int_{a+(1/n)}^{b+(1/n)} F(x) \, dx,$$
and so
$$\int_a^b f_n(x) \, dx = n \int_b^{b+(1/n)} F(x) \, dx - n \int_a^{a+(1/n)} F(x) \, dx.$$

The right hand side tends to $F(b) - F(a)$ in view of the continuity of F and Theorem 3.4.1.

11. As $\epsilon_n f \to |f|$ a.e. and $|\epsilon_n f| \leq |f|$, the Theorem of Dominated Convergence shows that
$$\int_a^b \epsilon_n f \to \int_a^b |f|.$$
But each integral
$$\int_a^b \epsilon_n f$$
is a finite sum of the form
$$\sum_{i=1}^r \pm (F(b_i) - F(a_i))$$
where (a_i, b_i) $(i = 1, 2, \ldots, r)$ are disjoint subintervals of $[a, b]$. These sums are bounded above by T and so
$$\int_a^b |f| \leq T.$$

12. *Lemma.* Suppose that $f(h)$ does not tend to l as $h \to 0$. Then there exists an $\epsilon > 0$ such that, for any integer $n \geq 1$, $|f(h_n) - l| \geq \epsilon$ for some h_n satisfying $0 < |h_n| < 1/n$. This sequence $\{h_n\}$ yields a contradiction.

By the Mean Value Theorem
$$\left|\frac{f(x, t+h) - f(x, t)}{h}\right| = \left|\frac{\partial}{\partial t} f(x, t + \theta h)\right| \leq g(x)$$
for any x, t. Now use the Theorem of Dominated Convergence and the above lemma.

14. For $x, t > 0$, $e^{tx} > tx$ and so $e^{-tx}/x < 1/(tx^2)$. The existence of $F(t)$ follows for all $t > 0$ by considering x in $(0, 1), (1, \infty)$ separately as in Ex. 5.1.6.

(i) For given $t > 0$ find c, d such that $0 < c < t < d$ and apply Ex. 13 to get
$$F'(t) = -\int_0^\infty e^{-tx} \sin x \, dx.$$

SOLUTIONS

It is a standard exercise in calculus (e.g. integrating by parts twice) to show that
$$\int_0^k e^{-tx}\sin x\,dx = 1/(1+t^2) - (\cos k + t\sin k)e^{-kt}/(1+t^2).$$
Let $k \to \infty$.

(ii) $\dfrac{d}{dt}(F(t)+\tan^{-1}t) = 0$ for all $t > 0$.

(iii) $F(n) = \displaystyle\int_0^\infty e^{-u}\dfrac{\sin(u/n)}{u}\,du.$ The integrand is dominated by e^{-u}/u on $(1,\infty)$. Thus $F(n) \to 0$ as $n \to \infty$. Hence $C = \tfrac{1}{2}\pi$.

(iv) According to (iii) $F(t) \to \tfrac{1}{2}\pi$ as $t \downarrow 0$. Secondly, for any $X > 0$,
$$\int_0^X e^{-x/n}\frac{\sin x}{x}\,dx \to \int_0^X \frac{\sin x}{x}\,dx \quad \text{as} \quad n \to \infty$$
by the Dominated Convergence Theorem. Thirdly,
$$\int_X^Y e^{-tx}\frac{\sin x}{x}\,dx = -\left[\frac{e^{-tx}\cos x}{x}\right]_X^Y - \int_X^Y \frac{(1+tx)e^{-tx}}{x^2}\cos x\,dx:$$
use $1+tx \leq e^{tx}$ and let $Y \to \infty$ to see that
$$\left|\int_X^\infty e^{-tx}\frac{\sin x}{x}\,dx\right| \leq 2/X \quad \text{for all} \quad t > 0.$$

Fourthly, we have already proved on p. 102 that $\displaystyle\int_0^X \frac{\sin x}{x}\,dx$ tends to a limit, l, say as $X \to \infty$ (cf. the lemma proved in Ex. 12).

Given any $\epsilon > 0$; choose X so that $2/X < \epsilon$ and also $\left|\displaystyle\int_0^X \frac{\sin x}{x}\,dx - l\right| < \epsilon$, then choose n so that $|F(1/n) - \tfrac{1}{2}\pi| < \epsilon$ and also
$$\left|\int_0^X e^{-x/n}\frac{\sin x}{x}\,dx - \int_0^X \frac{\sin x}{x}\,dx\right| < \epsilon.$$
Combining these gives $|l - \tfrac{1}{2}\pi| < 4\epsilon$. As ϵ is arbitrarily small we deduce finally that $l = \tfrac{1}{2}\pi$.

§ 6.1

2. By Ex. 5.2.8 and Theorem 1, $\text{mid}\{f,g,h\}$ is measurable; also
$$|\text{mid}\{f,g,h\}| \leq \max\{|g|,|h|\}$$
and so Proposition 1 applies.

3. If $|f| \leq K$ then $|f^2| \leq K|f|$ and Proposition 1 applies.

4. Let g be a positive measurable function and let $g^{\frac{1}{2}}$ denote the positive square root of g. For any positive K and any bounded interval I, $\min\{g, K\chi_I\}$

is integrable and so is the limit a.e. of a sequence $\{\phi_n\}$ of step functions. By suitable truncating we may arrange that ϕ_n is positive and is dominated by $K\chi_I$. Each $\phi_n^{\frac{1}{2}}$ is a step function and $\{\phi_n^{\frac{1}{2}}\}$ converges to min $\{g^{\frac{1}{2}}, K^{\frac{1}{2}}\chi_I\}$ a.e. By the Dominated Convergence Theorem, min $\{g^{\frac{1}{2}}, K^{\frac{1}{2}}\chi_I\}$ is integrable. Thus $g^{\frac{1}{2}}$ is measurable. [An alternative proof is available using Proposition 1 of the next section – which corresponds to the customary definition of a measurable function.]

Now take $g = f^2$ so that $g^{\frac{1}{2}} = |f|$.

For example, take $f(x) = 1/x$ for $x \geqslant 1$, $= 0$ otherwise. Then $f^2 \in L^1$ and $|f| = f \notin L^1$.

5. Use the first part of Ex. 4 and the inequalities $|f| \leqslant \sqrt{(f^2+g^2)}$, $|g| \leqslant \sqrt{(f^2+g^2)} \leqslant |f|+|g|$.

6. fg is measurable and $|fg| \leqslant \frac{1}{2}(f^2+g^2)$.

$$a^2 \int f^2 + 2ab \int fg + b^2 \int g^2 = \int (af+bg)^2 \geqslant 0$$

for all a, b in $\mathbf{R}$. The result follows from the elementary theory of quadratics.

7. Cf. the proof of the Dominated Convergence Theorem 5.2.1:

$$l_{nk} = \min\{f_n, \ldots, f_{n+k}\}$$

is measurable; $l_{nk} \to l_n$ as $k \to \infty$; $l_n \to \liminf f_n$ as $n \to \infty$. Similarly for $\limsup f_n$.

8. By Ex. 4.1.6 it is enough to prove the result for the characteristic function of a (non-empty) bounded open interval I. When $k = 1$ the result is clear from Fig. S. 6. More precisely, let I_n be the concentric closed interval of side $l(I) - 2/n$ (> 0 for large enough n), and let f_n be the continuous function defined by

$$f_n(x) = \frac{d(x, I^c)}{d(x, I^c) + d(x, I_n)}$$

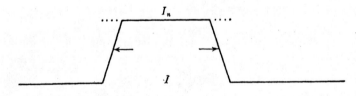

Fig. S. 6

(cf. Appendix, Exx. 5, 6). Then $f_n \to \chi_I$ as $n \to \infty$ and so χ_I is in the Baire class 1.

This construction generalises at once to $\mathbf{R}^k$.

9. The rationals may be written as the terms of a sequence $\{r_n\}$ without repeats. Let $S_n = \{r_1, \ldots, r_n\}$. Then the step functions χ_{S_n} belong to the Baire class 1 (by Ex. 8) and converge to χ_Q.

10. As usual define $\dfrac{\sin x}{x} = 1$ for $x = 0$.

Then
$$\int_0^X \frac{|\sin x|}{x} dx$$

exists as the integrand is continuous on $[0, X]$. Also
$$\int_0^\infty e^{-xy} dy = 1/x$$

for all $x > 0$. The given equation now holds by Tonelli's Theorem. By standard calculus (integration by parts twice)

$$\int_0^X e^{-xy} \sin x \, dx = -\left[e^{-xy} \frac{(\cos x + y \sin x)}{1+y^2} \right]_{x=0}^{x=X}$$
$$= \frac{1}{1+y^2} - e^{-Xy} \frac{(\cos X + y \sin X)}{1+y^2}.$$

Provided $X \geqslant 1$ and $y \geqslant 0$, we have $e^{Xy} \geqslant 1+y$ and
$$\left| e^{-Xy} \frac{(\cos X + y \sin X)}{1+y^2} \right| \leqslant \frac{1}{1+y^2}.$$

We may therefore let $X \to \infty$ through a sequence of values and apply the Dominated Convergence Theorem to obtain
$$\lim_{X \to \infty} \int_0^X \frac{\sin x}{x} dx = \int_0^\infty \frac{dy}{1+y^2} = \frac{\pi}{2}.$$

11. (i) $2xt - x^2 = t^2 - (t-x)^2$ attains the global maximum value t^2 when $x = t$. As the rationals are dense in $\mathbf{R}$ this value is approached arbitrarily closely for rational values of x.

(ii) Let $f_n = \max\{2r_k f - r_k^2 : 1 \leqslant k \leqslant n\}$. Then $\{f_n\}$ is a sequence of measurable functions which converges to f^2.

(iii) Immediate from (ii).

12. For the last part, $|\sin x| < 1$ except when $x = (k + \tfrac{1}{2})\pi$, $k \in \mathbf{Z}$.

§ 6.2

2. If $\{S_n\}$ is a decreasing sequence of sets of *finite* measure and $S = \bigcap S_n$, then S has finite measure $m(S) = \lim m(S_n)$. For the proof, let $T_n = S_1 \setminus S_n$ and apply Theorem 1 (ii).

This argument does not apply to the given example for which $m(S_n) = \infty$ for all n, and $m(S) = 0$.

3. Let $f = \chi_E$ where E is the non-measurable set described at the end of the section. If $I = [0,1]$ let $g = \chi_{E \times I}$, i.e. $g(x,y) = 1$ if $x \in E$, $y \in I$; $= 0$ otherwise.

4. (i) Let E, I be as in Ex. 3 and let $F = I \setminus E$. Then $f = \chi_E - \chi_F$ is non-measurable but $f^2 = \chi_I$.

(ii) Let $A = \{x: f(x) > 0\}$, $B = \{x: f(x) \leq 0\}$. Then A, B are measurable and $f = |f|(\chi_A - \chi_B)$ where $|f|$ is measurable by Ex. 6.1.4.

5. (i) Let $g = 1/f$. If $c > 0$, $\{x: g(x) \geq c\} = \{x: 0 < f(x) \leq 1/c\}$;
$$\{x: g(x) \geq 0\} = \{x: f(x) \geq 0\};$$
if $c < 0$, $\quad \{x: g(x) \geq c\} = \{x: f(x) \geq 0\} \cup \{x: f(x) \leq 1/c\}$.

(ii) The simple functions ϕ_n constructed in the proof of Proposition 2 all have 0 as one of the division points c_i and so $\phi_n(x) = 0$ whenever $f(x) = 0$. Thus $\{1/\phi_n\}$ converges to $1/f$ everywhere.

6. (i) If $c \leq d$, $f^{-1}\{[c,d]\} = A_c \backslash B_d$, $f^{-1}\{(c,d)\} = B_c \backslash A_d$, etc. On the other hand,
$$A_c = f^{-1}\{[c, \infty)\} = \bigcup_{n=1}^{\infty} f^{-1}\{[c, n)\}.$$

(ii) Let $\phi = c_1 \chi_{I_1} + \ldots + c_r \chi_{I_r}$, where $I_1, \ldots, I_r$ are disjoint bounded intervals. Then
$$\phi \circ f = c_1 \chi_{S_1} + \ldots + c_r \chi_{S_r},$$
where $S_i = f^{-1}(I_i)$ for $i = 1, \ldots, r$.

7. (i) If $\chi_{S_1}, \ldots, \chi_{S_r}$ are integrable then ϕ is integrable. On the other hand, if ϕ is integrable, then each $S_i = \{x: \phi(x) = c_i\}$ is measurable and $\chi_{S_i} \leq |c_i|^{-1}|\phi|$; thus χ_{S_i} is integrable for $i = 1, \ldots, r$. The given equation is obvious by linearity.

(ii) Apply the Monotone Convergence Theorem.

8. (i) Use Proposition 5.1.1 with $I_n = (-n, n)$ to show that $\int g = 2$. [The main point is that g is a *strictly positive integrable* function.]

(ii) The function $q: \mathbf{R}^2 \to \mathbf{R}$ defined by
$$q(x, y) = xy/(x+y) \quad \text{if} \quad x > 0, y > 0; = 0 \quad \text{otherwise},$$
is easily seen to be continuous. As f^+ and g are measurable it follows from Proposition 3 that h is measurable. But $0 \leq h < g$ and so h is integrable.

(iii) Let ϕ_n vanish outside the bounded interval I_n; then $g \chi_{I_n}$ is a step function, and so ψ_n is a step function. As
$$f^+ = gh/(g-h)$$
$\{\psi_n\}$ converges to f^+ a.e. Argue similarly for f^-.

(iv) Let $f = f^+ - f^-$ and use (iii).

9. All we need is a strictly positive function g in $L^1(\mathbf{R}^k)$; the rest of the argument is identical. If $x = (x_1, \ldots, x_k) \in \mathbf{R}^k$ let $\|x\| = \max\{|x_1|, \ldots, |x_k|\}$. Now define $g: \mathbf{R}^k \to \mathbf{R}$ by
$$g(x) = 1/2^n \quad \text{for} \quad n-1 \leq \|x\| < n \quad (n = 1, 2, \ldots)$$
and verify that $g \in L^1(\mathbf{R}^k)$. (The series $\Sigma((2n)^k/2^n)$ is convergent by the Ratio Test.)

SOLUTIONS   261

10. (i) Let J be a bounded interval in $\mathbf{R}^m$ with $m(J) > 0$. For any $\epsilon > 0$, we may cover N by a sequence $\{I_n\}$ of intervals in $\mathbf{R}^l$ for which $\Sigma m(I_n) < \epsilon/m(J)$. In this way $N \times J$ is covered by the sequence $\{I_n \times J\}$ for which $\Sigma m(I_n \times J) < \epsilon$, and so $N \times J$ is null. Now express $\mathbf{R}^m$ as the union of a sequence $\{J_n\}$ of bounded intervals and use the fact that each $N \times J_n$ is null.

(ii) First of all, let A, B have finite measures. Then there exist sequences $\{\phi_n\}$, $\{\psi_n\}$ of step functions such that $\phi_n \to \chi_A$ outside a null set S in $\mathbf{R}^l$ and $\psi_n \to \chi_B$ outside a null set T in $\mathbf{R}^m$. Let $\theta_n(x,y) = \phi_n(x)\psi_n(y)$ for all x in $\mathbf{R}^l$, y in $\mathbf{R}^m$; then θ_n is a step function on $\mathbf{R}^k$ and $\theta_n \to \chi_{A \times B}$ outside the set $(S \times \mathbf{R}^m) \cup (\mathbf{R}^l \times T)$ which is null by part (i). Thus $\chi_{A \times B}$ is measurable. By Tonelli's Theorem (extended in the obvious way to $\mathbf{R}^k$),

$$m(A \times B) = m(A)\, m(B)$$

in this case.

Now suppose only that A, B are measurable. Let $\{I_n\}$, $\{J_n\}$ be increasing sequences of intervals whose unions are $\mathbf{R}^l$, $\mathbf{R}^m$, respectively and apply the same argument to $(A \cap I_n)$ and $(B \cap J_n)$. We deduce that $A \times B$ is measurable and moreover

$$m(A \times B) = m(A)\, m(B)$$

will hold for all measurable sets A, B provided we interpret $\infty.\infty = \infty$ and $\infty.0 = 0.\infty = 0$ (the latter because these infinities are 'approached from below').

11. (i) $Q_{\chi_A} = \{(x,y): 0 \leq y \leq \chi_A(x)\} = A \times [0,1] \cup A^c \times \{0\}$.

(ii) Immediate by linearity.

(iii) Let $\{\phi_n\}$ be an increasing sequence of simple functions which converges everywhere to f. Then $P_{\phi_n} \uparrow P_f$.

(iv) $Q_f = \bigcap_{n=1}^{\infty} P_{f+(1/n)}$.

(v) Apply Ex. 7 and part (iii) above.

12. (i) Ignore x outside $[a,b]$. As f is increasing the sets S_i are intervals in this case (see the proof of Theorem 1 in the Appendix) and ϕ is a positive increasing step function. For convenience redefine ϕ, if necessary, so that $\phi(b) = f(b)$.

(ii) Let ϕ be a positive increasing step function on $[a,b]$: then there are points $a = x_0 \leq x_1 \leq \ldots \leq x_r = b$ such that

$$\phi = c_1 \chi_{I_1} + \ldots + c_r \chi_{I_r},$$

where I_i has end points x_{i-1}, x_i for $i = 1, \ldots, r$, and

$$\phi(a) = c_1 \leq c_2 \leq \ldots \leq c_r = \phi(b).$$

Let $$G(x) = \int_x^b g$$

for x in $[a,b]$. Then

$$\int_a^b \phi g = c_1(G(x_0)-G(x_1))+\ldots+c_r(G(x_{r-1})-G(x_r))$$
$$= c_1 G(x_0)+(c_2-c_1)G(x_1)+\ldots+(c_r-c_{r-1})G(x_{r-1})$$

(as $G(b)=0$). The values of the continuous function G lie between a minimum value m and a maximum value M. It follows that

$$m\phi(b) \leqslant \int_a^b \phi g \leqslant M\phi(b)$$

(because $0 \leqslant c_1 \leqslant c_2 \leqslant \ldots \leqslant c_r = \phi(b)$).

Let f be a positive increasing function on $[a,b]$. For any $\epsilon > 0$, construct ϕ as in part (i): then

$$mf(b)-\epsilon \int_a^b |g| \leqslant \int_a^b fg \leqslant Mf(b)+\epsilon \int_a^b |g|.$$

As ϵ is arbitrarily small $\quad mf(b) \leqslant \int_a^b fg \leqslant Mf(b)$

and so
$$\int_a^b fg = G(c)f(b)$$

for some c in $[a,b]$ by Theorems 3, 9 of the Appendix.

§ 6.3

2. Rotation through θ about 0. $\Delta_T = 1$. Replace θ by $-\theta$.

4. $\Delta_L = 1$.

6. As F is continous and increasing, T is continuous, strictly increasing and maps $[0,1]$ into $[0,2]$. Using Theorems 1, 2 of the Appendix, T maps $[0,1]$ *onto* $[0,2]$. The inverse mapping T^{-1} is strictly increasing and maps $[0,2]$ onto $[0,1]$. This leaves no room for jump discontinuities of T^{-1}, and so T^{-1} is continuous.

The function F is constant on the intervals which make up the complement $S^c = [0,1]\backslash S$ and T maps each of these intervals onto an interval of equal length. Thus $T(S^c)$ has measure 1 and so $T(S)$ has measure 1.

7. In the last six lines of the argument replace $(0,1)$ by F.

8. Let $T(S)$ be the set of measure 1 mentioned in Ex. 6. The construction of Ex. 7 may be adapted easily to the case where F is a subset of $[0,2]$ and so we may find a non-measurable subset B, say, of $T(S)$. Let $A = T^{-1}(B)$. Then A is null (because $A \subset S$) and $TA = B$ is non-measurable.

§ 6.4

1. Area of annulus = $\pi(r_2^2 - r_1^2)$ (see Fig. S. 7). Using additivity, the area of TI illustrated in Fig. 46, p. 147 is $\frac{1}{2}(r_2^2 - r_1^2)(\theta_2 - \theta_1)$ provided $\theta_2 - \theta_1$ is a rational multiple of 2π. The result follows as the rationals are dense in $\mathbf{R}$.

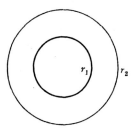

Fig. S. 7

2. Let $I = \{(r, \theta, z) : r_1 \leq r \leq r_2, \theta_1 \leq \theta \leq \theta_2, z_1 \leq z \leq z_2\}$. Using Ex. 1 and Fubini's Theorem the volume of the (compact) set TI is

$$\tfrac{1}{2}(r_2^2 - r_1^2)(\theta_2 - \theta_1)(z_2 - z_1).$$

3. $m(TI) = \int_0^{a \cos \alpha} \pi(a^2 - z^2 - z^2 \tan^2 \alpha)\, dz = \pi a^3 (\cos \alpha - \tfrac{1}{3} \sec^2 \alpha \cos^3 \alpha)$
$$= \tfrac{2}{3}\pi a^3 \cos \alpha.$$

This volume is unaltered if the boundary points are included. Let J be defined by $r_1 \leq r \leq r_2$, $\theta_1 \leq \theta \leq \theta_2$, $\phi_1 \leq \phi \leq \phi_2$. By additivity, the volume of TJ is
$$\tfrac{1}{3}(r_2^3 - r_1^3)(\cos \theta_1 - \cos \theta_2)(\phi_2 - \phi_1)$$

provided $\phi_2 - \phi_1$ is a rational multiple of 2π. Argue as in Ex. 1 that the rationals are dense in $\mathbf{R}$. Thus

$$m(TI) = \int_{r_1}^{r_2} \int_{\theta_1}^{\theta_2} \int_{\phi_1}^{\phi_2} r^2 \sin \theta \, dr \, d\theta \, d\phi.$$

4. The lines $x = 0$, $y = 0$ have zero area and so may be omitted from I without altering the value of the integral. Using the polar transformation of Ex. 1 and Theorem 1 (or alternatively Theorem 4) the given integral equals

$$\int_0^{\pi/2} \left(\int_0^{\infty} e^{-r^2} r \, dr \right) d\theta = \pi/4 \int_0^{\infty} e^{-u} du \quad \text{(substitute } u = r^2\text{)}$$
$$= \pi/4.$$

Now use Ex. 5.1.4.

5. Assume to begin with that V has finite measure. Thus $h \in L^1(U)$ and $m(V) = \int_U h$. If TE is measurable then TE has finite measure and

equation (3) of Theorem 1 gives $m(TE) = \int_E h$. Thus $h\chi_E$ is integrable and so
$$E \cap S = \{x : h(x)\chi_E(x) > 0\}$$
is measurable.

On the other hand if $E \cap S$ is measurable, $h \in L^1(E \cap S)$. As h vanishes outside S, $h \in L^1(E)$ and equation (3) shows that TE has finite measure
$$\int_E h.$$

Let J_n be the open cube in $\mathbf{R}^k$ of side $2n$, centre 0, let $A_n = T^{-1}J_n$ and let $E_n = E \cap A_n$. The above argument applies to the open set A_n in place of U and shows that TE_n is measurable if and only if $E_n \cap S$ is measurable. If we now let $n \to \infty$ this establishes the general case where V may have infinite measure.

6. Given $p \in U$; as U is open there exists $r > 0$ such that any closed cube of side $< r$ containing p is contained in U. Suppose that
$$\frac{m(TC)}{m(C)} \to l \quad \text{as} \quad C \to p,$$
where C is restricted to open cubes containing p (with $\overline{C} \subset U$). To any $\epsilon > 0$, there is a δ in $(0, r)$ such that
$$l - \epsilon \leqslant \frac{m(TC)}{m(C)} \leqslant l + \epsilon$$
for any open cube C of side $< \delta$ containing p. Now let Q be any closed cube of side $< \delta$ containing p (with $m(Q) > 0$). Construct a decreasing sequence $\{C_n\}$ of open cubes concentric with Q all of side $< \delta$ and whose intersection is Q. As $\delta < r$, each $\overline{C}_n \subset U$. By the Monotone Convergence Theorem
$$m(C_n) \to m(Q) \quad \text{and} \quad m(TC_n) \to m(TQ).$$
Thus
$$l - \epsilon \leqslant \frac{m(TQ)}{m(Q)} \leqslant l + \epsilon.$$
This shows that
$$\frac{m(TQ)}{m(Q)} \to l \quad \text{as} \quad Q \to p,$$
where Q is restricted to closed cubes containing p (with $m(Q) > 0$).

An almost identical argument with an increasing sequence of concentric closed cubes shows that the condition for closed cubes implies the condition for open cubes. If C is an open cube containing p whose closure $\overline{C}$ is contained in U, and C' is any cube satisfying $C \subset C' \subset \overline{C}$ then
$$m(C) = m(C') = m(\overline{C})$$
and so
$$\frac{m(TC)}{m(C)} \leqslant \frac{m(TC')}{m(C')} \leqslant \frac{m(T\overline{C})}{m(\overline{C})}.$$

The equivalence of our given definition and the restricted definitions (i), (iii) follows at once from these inequalities by the customary squeezing argument.

A similar argument shows the equivalence with (ii).

SOLUTIONS

7. Let $L(x) = y$, i.e. $y_i = c_{i1}x_1 + \ldots + c_{ik}x_k$. Thus
$$|y_i| \leq (|c_{i1}| + \ldots + |c_{ik}|)\|x\| \leq K\|x\| \quad \text{for all } i.$$

8. Using Ex. 7 $\|T(x+h) - T(x)\| \leq (K_x + \epsilon)\|h\| \to 0$ as $\|h\| \to 0$.

9. Let x approach 0 along the parabola $x_1 = x_2^2$ on which $f(x_1, x_2) = \frac{1}{2}$ except when $x_1 = x_2 = 0$.

10. By the Mean Value Theorem
$$f(x_1+h_1, x_2+h_2) - f(x_1, x_2+h_2) = h_1 D_1 f(x_1 + \theta_1 h_1, x_2 + h_2),$$
$$f(x_1, x_2+h_2) - f(x_1, x_2) = h_2 D_2 f(x_1, x_2 + \theta_2 h_2),$$
where $0 < \theta_1, \theta_2 < 1$. As $D_1 f, D_2 f$ are continuous and as $|h_1|, |h_2| \leq \|h\|$, it follows that
$$f(x+h) - f(x) = h_1 D_1 f(x) + h_2 D_2 f(x) + o(h).$$

The second part is immediate by considering the components T_1, T_2.

12. The Jacobian matrices are

$$\begin{pmatrix} \cos\theta & -r\sin\theta \\ \sin\theta & r\cos\theta \end{pmatrix}, \quad \begin{pmatrix} \cos\theta & -r\sin\theta & 0 \\ \sin\theta & r\cos\theta & 0 \\ 0 & 0 & 1 \end{pmatrix},$$

$$\begin{pmatrix} \sin\theta\cos\phi & r\cos\theta\cos\phi & -r\sin\theta\sin\phi \\ \sin\theta\sin\phi & r\cos\theta\sin\phi & r\sin\theta\cos\phi \\ \cos\theta & -r\sin\theta & 0 \end{pmatrix}, \quad \text{respectively.}$$

The corresponding density functions are $r, r, r^2 \sin\theta$.

13. (i) The determinant of a matrix is a polynomial in the coefficients of the matrix. In this case the coefficients of the matrix are themselves continuous functions and so Δ_T is continuous. (From the criterion for continuity given in Proposition 3 of the Appendix, it is easy to verify that the composite of two continuous functions is continuous.)

(ii) Let $p \in U$ and suppose that the partial derivatives $D_j T_i$ are continuous at p. Given any $\epsilon > 0$; there exists $\delta > 0$ such that

$$x \in U \quad \text{and} \quad |D_j T_i(x) - D_j T_i(p)| < \epsilon/k \quad \text{for} \quad \|x - p\| < \delta$$
$$(i, j = 1, \ldots, k).$$

In the light of Ex. 7 this implies that
$$\|L_x(h) - L_p(h)\| \leq \epsilon \|h\| \qquad (*)$$

for all h in $\mathbf{R}^k$ provided $\|x - p\| < \delta$. This inequality is analogous to inequality (6) in the proof of Proposition 1.

Let C be the unit cube $\|x\| \leq \frac{1}{2}$. Recall that $\Delta_T(x) = m(L_x C)$. Now argue as in Proposition 1. First of all, we may assume without loss of generality that L_p has diagonal matrix so that $L_p C$ is an interval.

Case 1. If $\Delta_T(p) = 0$ then $m(L_p C) = 0$ and the inequality $(*)$ shows that $m(L_x C) \to 0$ as $\|x - p\| \to 0$.

Case 2. If $\Delta_T(p) \neq 0$ then L_p has an inverse and we may assume without loss of generality that L_p is the identity mapping. The inequality (∗) simplifies to
$$\|L_x(h) - h\| \leq \epsilon \|h\|$$
for all h in $\mathbf{R}^k$ provided $\|x-p\| < \delta$. The argument of Proposition 1 now shows that $m(L_x C) \to 1$ as $\|x-p\| \to 0$. (At one stage the argument is simplified by observing that $L_x C$ and C_- have the point 0 in common.)

Thus in both cases we have proved that
$$\Delta_T(x) \to \Delta_T(p) \quad \text{as} \quad \|x-p\| \to 0.$$

14. $T(x,y) = (s,t) = (xy, y/x)$. $B = TA$ is given by the inequalities $0 < s < 3$, $1 < t < 2$. (Draw a diagram!)
$$\Delta_T = \begin{vmatrix} y & x \\ -y/x^2 & 1/x \end{vmatrix} = 2y/x = 2t.$$

By Theorem 1, Corollary $\Delta_{T^{-1}} = 1/2t$. As $y^2 = st$, the required integral is
$$\frac{1}{2}\int_B s\, d(s,t) = \frac{1}{2}\int_1^2 \left(\int_0^3 s\, ds \right) dt = \tfrac{9}{4}.$$

§ 7.1

1. (i) By a simple inductive argument, $2^n > n$ for $n \geq 1$. Given $\epsilon > 0$; by Archimedes' Axiom there is an integer $N > 1/\epsilon$ and so
$$2^{-n} < 1/n < \epsilon \quad \text{for} \quad n \geq N.$$

(ii) Let $c_1 = \tfrac{1}{2}(s_1 + K)$. One of the intervals $[s_1, c_1]$, $[c_1, K]$ contains s_n for infinitely many values of n: call this interval I_1. Let I_2 be a closed half of I_1 that contains s_n for infinitely many values of n, and so on. Let $N(1)$ be the first integer n for which $s_n \in I_1$, let $N(2)$ be the first integer $n > N(1)$ for which $s_n \in I_2$, and so on. As $s_{N(k)} \in I_k$ for all $k \geq 1$ and $l(I_k) \to 0$ by part (i), it is clear that the subsequence $\{s_{N(k)}\}$ satisfies the Cauchy condition.

(iii) This subsequence converges to a limit s, say, in $\mathbf{R}$. It is easy now to verify that $s_n \to s$.

§ 7.2

2. $f(x) = 1 - c + cx - x^c$,
$f'(x) = c(1 - x^{c-1}) \quad (x > 0);$

thus
$$f'(x) < 0 \quad \text{for} \quad 0 < x < 1;$$
$$= 0 \quad \text{for} \quad x = 1;$$
$$> 0 \quad \text{for} \quad 1 < x.$$

By the Mean Value Theorem f is strictly decreasing in $[0,1]$, strictly increasing in $[1, \infty)$ and attains the global minimum value 0 when $x = 1$. The deduction of Lemma 1 is now straightforward.

SOLUTIONS

3. $f_n(x) = n$ for $0 < x < 1/n$; $= 0$ otherwise. Let $m < n$. Then
$$\|f_m - f_n\| = 2(1 - m/n)$$
and so $\{f_n\}$ does not satisfy the Cauchy condition.

4. Use the inequalities $|x_i - y_i| \leq \|x - y\| \leq \Sigma |x_i - y_i|$.

5. L^s is a linear space as for L^p ($p \geq 1$). The case $s = 1$ is obvious; suppose that $0 < s < 1$ and let t be the conjugate index defined by $(1/s) + (1/t) = 1$. Let f, h be positive measurable functions for which $f^s \in L^1$, $h^t \in L^1$. As $t < 0$ the last condition implies that $h > 0$ a.e. Let $p = 1/s$ and $(1/p) + (1/q) = 1$; then $q = (1-s)^{-1}$. If $F = f^s h^s$, $H = h^{-s}$, then $H^q = h^t \in L^1$ and $F^p = fh$. In the first place assume that $fh \in L^1$ and apply Hölder's Inequality to the positive functions F, H; then

$$\int FH \leq \left(\int F^p\right)^{1/p} \left(\int H^q\right)^{1/q}$$

$$\int f^s \leq \left(\int fh\right)^{1/p} \left(\int h^t\right)^{1/q}$$

which yields
$$\left(\int fh\right) \geq \left(\int f^s\right)^{1/s} \left(\int h^t\right)^{1/t}$$

provided $\int h^t \neq 0$. Even if $fh \notin L^1$ we may regard this inequality as true in the sense that the left hand side is infinite.

Now suppose that f, g are positive functions in L^s and consider
$$(f+g)^s = f(f+g)^{s-1} + g(f+g)^{s-1}.$$

As
$$(s-1)t = s; \quad (f+g)^{(s-1)t} = (f+g)^s \in L^1;$$

let
$$B = \int (f+g)^s.$$

If $B = 0$, $f = g = 0$ a.e.; assume therefore that $B \neq 0$. The above inequality gives
$$B \geq (\|f\| + \|g\|) B^{1/t},$$
whence
$$\|f + g\| = B^{1/s} \geq \|f\| + \|g\|.$$

§ 7.3

3. Let A, B be two disjoint intervals on $\mathbf{R}$ of length 1 and $f = \chi_A$, $g = \chi_B$.

4.
$$f(x) = (1 + x^{1/p})^p + (1 - x^{1/p})^p,$$
$$f'(x) = (x^{-1/p} + 1)^{p-1} - (x^{-1/p} - 1)^{p-1},$$
$$f''(x) = -\frac{(p-1)}{p} x^{-(1+1/p)} \{(x^{-1/p} + 1)^{p-2} - (x^{-1/p} - 1)^{p-2}\}.$$

(i) As $p \geq 2$, $f''(x) \leq 0$; thus f is a 'concave' function and the graph of f lies below the tangent at $(c, f(c))$, i.e.

$$f(x) \leq f(c) + (x-c)f'(c)$$

for $0 < x, c < 1$. (Formally, the Mean Value Theorem leads to Taylor's expansion $f(x) = f(c) + (x-c)f'(c) + \tfrac{1}{2}(x-c)^2 f''(z)$ where z is between x and c.) As $c \uparrow 1$ this gives

$$f(x) \leq 2^p + (x-1)2^{p-1} = 2^{p-1}(x+1).$$

The substitution $c = x^q$ gives $f(x) \leq 2(1 + x^{q/p})^{p/q}$ (after a simple reduction).

(ii) If $1 < p \leq 2$, $f''(x) \geq 0$ and all the subsequent inequalities are reversed.

5. (i) The inequalities for a, b are obviously true if $a = 0$ or $b = 0$ or $a = b$; also, they are unaltered if the sign of a or b is changed or if a, b are interchanged. We may therefore assume without loss of generality that $0 < b < a$ and substitute $x = b^p/a^p$ in the inequalities of Ex. 4 (i). This gives

$$(a+b)^p + (a-b)^p \leq 2^{p-1}(a^p + b^p),$$

$$(a+b)^p + (a-b)^p \leq 2(a^q + b^q)^{p/q}.$$

The other two inequalities are immediate if we substitute $a+b = c$, $a-b = d$.

Clarkson's Inequalities follow by integrating. The first is obvious; the second (or third) requires Minkowski's Inequality for L^s with $s = p/q \geq 1$:

$$\int |f+g|^p + \int |f-g|^p \leq 2 \int \{|f|^q + |g|^q\}^{p/q} \leq 2 \left\{ \left(\int |f|^p \right)^{q/p} + \left(\int |g|^p \right)^{q/p} \right\}^{p/q}$$

whence
$$\|f+g\|^p + \|f-g\|^p \leq 2\{\|f\|^q + \|g\|^q\}^{p/q}.$$

The third inequality follows by the substitution $f+g = F$, $f-g = G$.

(ii) The only awkward point is taken care of by Ex. 7.2.5.

6. The 'garden trellis' argument given for Theorem 1 only requires a 'parallelogram inequality'. When $p \geq 2$ use

$$\|f+g\|^p + \|f-g\|^p \leq 2^{p-1}\{\|f\|^p + \|g\|^p\}$$

and when $1 < p \leq 2$ use

$$\|f+g\|^q + \|f-g\|^q \leq 2\{\|f\|^p + \|g\|^p\}^{q/p}.$$

In the case of L^1 let A, B be disjoint subsets of $\mathbf{R}$ each of length 1, let $f = \chi_A$, $g = \chi_B$, let $h_t = (1-t)f + tg$ and let $S = \{h_t : t \in \mathbf{R}\}$. If $0 \leq t \leq 1$, $\|h_t\| = 1$; if $t > 1$, $\|h_t\| = 2t - 1 > 1$; and if $t < 0$, $\|h_t\| = 1 - 2t > 1$. Thus the closed convex set S has no unique element of smallest norm.

7. $\qquad \|Pf\| \leq \|f\|, \quad \|Qf\| \leq \|f\|.$

§ 7.4

1. Let $a = (a_1, a_2, a_3)$; then $F(x)$ is the scalar product $a.x$. Now show, as in Theorem 2, that $\|F\| = |a|$.
When $|a| = 1$, $a.(x-a) = 0$ for any x in the plane $a.x = 1$.

2. Let S be the given sphere and H the given hyperplane. By Theorem 7.3.1, there is a unique point h of H nearest the origin.
As $\|F\| = 1$, Schwarz' Inequality shows that $1 = F(f) \leq \|f\|$ for any f in H, and in particular $1 \leq \|h\|$. To any ϵ in $(0, 1)$, there is a g in L^2 such that $|F(g)| > (1-\epsilon)\|g\|$. For a suitable non-zero scalar t,

$$f = tg \in H \quad \text{and} \quad 1 = F(f) > (1-\epsilon)\|f\| \geq (1-\epsilon)\|h\|$$

(by the minimum property of $\|h\|$). As ϵ is arbitrary in $(0, 1)$ we deduce that $1 \geq \|h\|$. Thus $\|h\| = 1$ and $h \in S \cap H$. There is only one such point (to within a null function) as it is a point of H at the minimum distance 1 from the origin.

3. If $F(m) = 0$ then $h + tm \in H$ and so $\|h + tm\|^2$ attains the minimum value 1 when $t = 0$.

$$\|h + tm\|^2 = \|h\|^2 + 2t(h, m) + t^2\|m\|^2.$$

Differentiating with respect to t and substituting $t = 0$ gives $(h, m) = 0$.
Let $F(f) = s$. Then $F(f - sh) = 0$ and so $(h, f - sh) = 0$, i.e.

$$(h, f) = s(h, h) = s = F(f).$$

4. $\quad \phi_f(t) - \phi_f(0) = \|h + tf\| - \|h\| \geq F(h + tf) - F(h) = tF(f).$

Divide by $t \neq 0$; let t tend to zero from below and above and deduce that

$$F(f) \leq \phi_f'(0) \leq F(f).$$

As $\phi_f(t) = \{\|h\|^2 + 2t(h, f) + t^2\|f\|^2\}^{\frac{1}{2}}$, $\phi_f'(0)$ does in fact exist and equals (h, f).

5. $\operatorname{sgn} h = h/|h|$ is measurable. $|g|^q = |h|^{q(p-1)} = |h|^p$ and so $g \in L^q$. The other relations are immediate.

6. By Hölder's Inequality $\|F\| \leq \|g\|$. But $F(h) = \|g\|\|h\|$ (Ex. 5) and so $\|F\| \geq \|g\|$.
Let H be the given hyperplane. $F(h) = \int |g|^q$ (Ex. 5) $= \|g\|^q$. By Hölder's Inequality, for any f in H, $|F(f)| = \|F\|^q \leq \|F\|\|f\|$ and so $\|f\| \geq \|F\|^{q-1} = \|h\|$ (Ex. 5). (This is trivially true if $\|F\| = 0$.)

7. Just as for Ex. 2, except that Hölder replaces Schwarz.

8. (i) If $F(m) = 0$, $h + tm \in H$.
(ii) $\psi'(t) = \frac{1}{2}p(a^2 + 2tab + t^2b^2)^{\frac{1}{2}p-1}(2ab + 2tb^2) = p|a + tb|^{p-2}(a + tb)b$

provided $a+tb \neq 0$. Suppose that $a+tb = 0$. The case $a = b = 0$ is trivial, so we assume that $b \neq 0$, $t = -a/b$. Then

$$\frac{|a+sb|^p - |a+tb|^p}{s-t} = b\frac{|a+sb|^p}{a+sb} \to 0 \quad \text{as} \quad s \to t$$

(because $p > 1$).

(iii) The derivative of the integrand is $p|h+tm|^{p-2}(h+tm)m$ which is dominated by $p(|h|+|m|)^{p-1}|m|$ for $|t| \leq 1$. This last function is integrable (by Hölder's Inequality); thus we may differentiate as in Ex. 5.2.13 and substitute $t = 0$ to obtain $\int |h|^{p-2}hm = 0$, i.e. $\int gm = 0$.

(iv) Let $F(f) = s$. Then $F(f-sh) = 0$ and so $\int g(f-sh) = 0$, i.e.

$$\int gf = s \int gh = s = F(f).$$

9. At the last stage $\phi_f(t) = (\int |h+tf|^p)^{1/p}$ and so

$$\phi_f'(0) = \frac{1}{p}\|h\|^{1-p} p \int gf = \int gf.$$

10. By Ex. 6, θ is a norm-preserving linear mapping of L^q into the dual space of L^p. For any $F \neq 0$ we may divide by $\|F\|$ and apply Ex. 8 to see that θ is *onto* the dual space of L^p.

§ 7.5

4. Let $\{f_1, \ldots, f_r\}$ be a basis of $M \cap N$. Use Ex. 3 to find bases

$$\{f_1, \ldots, f_r, g_{r+1}, \ldots, g_m\} \quad \text{of } M \text{ and} \quad \{f_1, \ldots, f_r, h_{r+1}, \ldots, h_n\}$$

of N. Verify that the f's, g's, h's together form a basis of $M+N$.

5. $|\phi(x-y)| = \|(x_1-y_1)f_1 + \ldots + (x_m-y_m)f_m\|$

$\leq |x_1-y_1| + \ldots + |x_m-y_m| \leq m|x-y| \to 0 \quad \text{as} \quad |x-y| \to 0$.

Thus ϕ is continuous.

The image of the unit sphere S is a compact set C of positive real numbers. As $f_1, \ldots, f_r$ are linearly independent, C does not contain 0 and so by Theorem 9 of the Appendix, there exists $a > 0$ such that $\phi(x) \geq a$ for all x on S. Hence $\phi(x) \geq a|x|$ for all x in $\mathbf{R}^m$ by the homogeneity of ϕ. The inequalities

$$a|x-y| \leq \phi(x-y) \leq m|x-y|$$

now follow.

6. (i) This follows at once from the completeness of $\mathbf{R}$ if we note the inequalities $|x_i - y_i| \leq |x-y| \leq |x_1-y_1| + \ldots + |x_m-y_m|$.

(ii) Immediate from the inequalities of Ex. 5.

(iii) If $f_n \in M$ ($n = 1, 2, \ldots$) and $\|f_n - f\| \to 0$ ($f \in L^2$), then $\|f_m - f_n\| \to 0$ as $m, n \to \infty$. Thus $\{f_n\}$ converges to a limit in M and this limit must be f.

7. $\quad P_0(x) = 1, \quad P_1(x) = x, \quad P_2(x) = \tfrac{1}{2}(3x^2 - 1).$
$\quad L_0(x) = 1, \quad L_1(x) = -x+1, \quad L_2(x) = x^2 - 4x + 2.$
$\quad H_0(x) = 1, \quad H_1(x) = 2x, \quad H_2(x) = 4x^2 - 2.$

8. At each stage use successive integration by parts.

§ 7.6

3. (i) Consider $f = \chi_{(0,\pi)} - \chi_{(-\pi,0)}$.
(ii) Consider $|x|$.

5. (i) $$h_n(t) = \frac{1}{\pi}\int_I D_n(x-t)\,dx = \frac{1}{\pi}\int_J D_n(x)\,dx,$$

where J is an interval of length $\leq 2\pi$. As D_n has period 2π, $|h_n(t)| < 6/\pi$ by Lemma 1.

(ii) Let $$h_n(t) = \tfrac{1}{2}c_0 + \sum_{k=1}^{n}(c_k \cos kt + d_k \sin kt).$$
By equations (1),
$$\frac{1}{\pi}\int_{-\pi}^{\pi} h_n(t)f(t)\,dt = \tfrac{1}{2}c_0 a_0 + \sum_{k=1}^{n}(c_k a_k + d_k b_k).$$
Suppose that $0 \leq x \leq \pi$ and let $I = [0, x]$. Then
$$c_k = \frac{1}{\pi}\int_0^x \cos kt\,dt = x/\pi \quad \text{if } k = 0,$$
$$= \sin kx / k\pi \quad \text{if } k \geq 1,$$
$$d_k = \frac{1}{\pi}\int_0^x \sin kt\,dt = \frac{1 - \cos kx}{k\pi} \quad \text{if } k \geq 1.$$

Now $h_n \to \chi_I$ a.e. by Proposition 3 and $|h_n f| \leq (6/\pi)|f|$ by part (i). Thus
$$\int_{-\pi}^{\pi} h_n f \to \int_I f$$
(Dominated Convergence Theorem) and this gives the result for x in $[0, \pi]$. It extends at once to x in $(-\pi, 0)$ and then by the periodicity of f to all x in $\mathbf{R}$.

(iii) The sequence $\left\{\int_{-\pi}^{\pi} h_n f\right\}$ converges to $F(x)$ and is bounded above by
$$\frac{6}{\pi}\int_{-\pi}^{\pi}|f|.$$
The Bounded Convergence Theorem gives
$$\sum_{k=1}^{\infty}\frac{b_k}{k} = \frac{1}{2\pi}\int_{-\pi}^{\pi} F(x)\,dx.$$

(It is noteworthy that the series $\Sigma a_k/k$ need not be convergent: see the standard text on Fourier series by A. Zygmund [29].)

7. Bessel's equation gives
$$\frac{1}{\pi}\int_{-\pi}^{\pi} f(x)^2\,dx = \tfrac{1}{2}a_0^2 + \sum_{k=1}^{\infty}(a_k^2+b_k^2).$$
Now use the series in 4(ii) and the series obtained by integrating it once term by term (Ex. 5).

8. f and $\chi_{[0,x]}$ both belong to L^2. Use Parseval's equation p. 199 to evaluate $\int f \chi_{[0,x]}$.

9. $$F(x+2\pi) = F(x) + \int_x^{x+2\pi} f = F(x) + \int_{-\pi}^{\pi} f = F(x) + \pi a_0$$
(using the periodicity of f). The periodicity of G is now immediate.

The Fourier coefficients of G are $A_k = b_k/k$, $B_k = -a_k/k$ (integrate by parts), and the constant term is C. Jordan's Theorem now gives Ex. 5 (iii).

10. Apply the Theorem of Bounded Convergence to $\{(s_n-\phi)^2\}$ which converges to zero a.e.

11. $$u(x) = \frac{x - 2\sin\tfrac{1}{2}x}{2x\sin\tfrac{1}{2}x} \to 0 \quad \text{as} \quad x \to 0$$
(differentiating numerator and denominator twice) and u is continuous at all points of $(0,\pi]$.

12. (i) $u\chi_{[0,\pi]} \in L^1$.
(ii) Hence, from (6), (7)
$$\int_0^\pi \frac{\sin(n+\tfrac{1}{2})t}{t}\,dt \to \tfrac{1}{2}\pi.$$
Substitute $x = (n+\tfrac{1}{2})t$.
(iii) $X \in ((n-\tfrac{1}{2})\pi, (n+\tfrac{1}{2})\pi]$ for some integer n and
$$\left|\int_{(n-\tfrac{1}{2})\pi}^{(n+\tfrac{1}{2})\pi} \frac{\sin x}{x}\,dx\right| \leq 1/(n-\tfrac{1}{2}).$$

13. $v(x) \to 1$ as $x \to 0$. $v'(x) = w(x)/x^2$ where $w(x) = x\cos x - \sin x$. $w'(x) = -x\sin x < 0$ for $0 < x < \pi$. By the Mean Value Theorem
$$w(x) - w(0) = w(x) < 0 \quad \text{for} \quad 0 < x < \pi$$
and hence v is strictly decreasing on $[0,\pi]$.

14. (i) There exists r in $(0,\pi]$ such that
$$\left|\frac{f(x+t)-f(x)}{t} - f'(x)\right| < 1 \quad \text{for} \quad 0 < |t| < r.$$
(ii) By (i), $\int_0^r \dfrac{g(x,t)}{t}\,dt$ exists.

SOLUTIONS

(iii) Apply the Riemann–Lebesgue Lemma (Theorem 1) and the Localisation Principle (Proposition 2).

15. (ii) $G'(x) = v(x)$.

(iii) $$\left| \int_{(k-1)\pi}^{k\pi} \frac{\sin t}{t} dt \right| = \int_0^{\pi} \frac{\sin x}{x + (k-1)\pi} dx \leq 1/(k-1).$$

(v) $G(x)$ is positive and has (global) maximum value $G(\pi)$.

17. The given area equals
$$\int_0^{\pi} \frac{\sin x}{\sin\left(\frac{x + (k-1)\pi}{2n+1}\right)} \frac{dx}{(2n+1)}.$$

For x in $[0, \pi]$
$$\sin\left(\frac{x + (k-1)\pi}{2n+1}\right)$$

increases as k increases from 1 to $n+1$. (A little care is needed for the last value, as the interval $[n\pi/(2n+1), (n+1)\pi/(2n+1)]$ straddles $\pi/2$.) It follows that the restriction of G to $[0, \pi]$ is positive and has global maximum at the point x_1.

19. (ii) Ex. 1.

(iii) $s_n'(x) = (4/\pi)(\cos x + \cos 3x + \ldots + \cos(2n-1)x)$. Multiply by $\sin x$ and use the identity $2 \sin x \cos(2k-1)x = \sin 2kx - \sin 2(k-1)x$; this gives
$$s_n'(x) = \frac{2}{\pi} \frac{\sin 2nx}{\sin x}.$$

(iv) $s_n'(x) = 0$ where $\sin 2nx = 0$ $(x \neq 0)$.

(v) Argue as in Exx. 15, 17.

(vi) $$s_n(\pi/2n) = \frac{2}{\pi} \int_0^{\pi} \frac{\sin x}{x} \frac{(x/2n)}{\sin(x/2n)} dx \downarrow L.$$

20. Use equation (10) and the positiveness of F_n; then use equation (14) and the periodicity of F_n.

Examples A–H are left as a challenge!

§ 7.7

1. The verification that l^p is a linear space and the proofs of Hölder's and Minkowski's Inequalities are almost the same as for L^p. The completeness of l^p may be proved as follows.

Let $x_1, x_2, \ldots$ be a Cauchy sequence in l^p. Thus
$$\sum_{r=1}^{\infty} |x_{mr} - x_{nr}|^p \to 0 \quad \text{as} \quad m, n \to \infty.$$

In particular, for each r, $|x_{mr}-x_{nr}| \to 0$ as $m, n \to \infty$ and so, by the completeness of the metric space **R**, $x_{mr} \to s_r$, say, as $m \to \infty$. Given any $\epsilon > 0$; there exists N such that

$$\sum_{r=1}^{R} |x_{mr}-x_{nr}|^p < \epsilon^p \quad \text{for} \quad m, n \geqslant N \quad \text{and any} \quad R \geqslant 1.$$

Let $m \to \infty$ in this (finite) sum:

$$\sum_{r=1}^{R} |s_r-x_{nr}|^p \leqslant \epsilon^p \quad \text{for} \quad n \geqslant N \quad \text{and any} \quad R \geqslant 1.$$

Let $R \to \infty$ and deduce that $s-x_n \in l^p$ and $\|s-x_n\| \leqslant \epsilon$ for $n \geqslant N$. Thus $s \in l^p$ and $\|s-x_n\| \to 0$ as $n \to \infty$.

[In the second volume we shall see that l^p is a space $L^p(\mu)$, where μ is the 'counting measure' on $\{1, 2, \ldots\}$. The completeness of l^p may thus be proved by an exact repetition of the argument in §7.2.]

2. Ex. 7.5.6.

3. To establish completeness, let $\{f_n\}$ be a Cauchy sequence in $C[0,1]$: to any $\epsilon > 0$, there exists N such that $|f_m(x)-f_n(x)| < \epsilon$ for all x in $[0,1]$ and all $m, n \geqslant N$. For each x in $[0,1]$ Cauchy's General Principle of Convergence ensures that $f_m(x) \to f(x)$, say, as $m \to \infty$. Let $m \to \infty$ in the above inequality; thus

$$|f(x)-f_n(x)| \leqslant \epsilon \quad \text{for all } x \text{ in} \quad [0,1] \quad \text{and all} \quad n \geqslant N.$$

It only remains to prove that f is *continuous* on $[0,1]$ as this inequality then shows that $\|f-f_n\| \to 0$ as $n \to \infty$. For any point p in $[0,1]$, the function f_N is continuous at p and so we may find $\delta > 0$ such that

$$|f_N(x)-f_N(p)| < \epsilon \quad \text{for all } x \text{ in} \quad [0,1] \quad \text{satisfying} \quad |x-p| < \delta.$$

Then

$$|f(x)-f(p)| \leqslant |f(x)-f_N(x)| + |f_N(x)-f_N(p)| + |f_N(p)-f(p)|$$
$$< 3\epsilon \quad \text{for all } x \text{ in } [0,1] \quad \text{satisfying} \quad |x-p| < \delta.$$

This establishes the continuity of f.

4. See Ex. 2.

5. Let S be an orthonormal set in L^2. Suppose that S is complete. If f is orthogonal to every element of S then $f = 0$ (strictly, f is null) and so S is maximal. Suppose that S is incomplete. By the Projection Theorem there exists a non-zero element in $S^\perp$ and so S is not maximal.

Appendix

1. $\bar{A}$ is the smallest closed set containing A.

2. (i) p is an adherent point of A if and only if $p \notin \text{ext } A$.

SOLUTIONS

(ii) An adherent point of A that does not belong to A is a limit point of A (and any limit point is certainly adherent).

5. $d(x, A) = 0 \Leftrightarrow$ every open ball centre x contains points of A
$\Leftrightarrow x$ is an adherent point of A.

6. (i) $d(x, A) \leq d(x, a) \leq d(x, y) + d(y, a)$ for any a in A. Thus
$$d(x, A) - d(x, y) \leq d(y, A), \quad \text{i.e.,} \quad d(x, A) - d(y, A) \leq d(x, y).$$
Interchange the roles of x, y and deduce the given inequality.

(ii) $|f(x) - f(y)| \leq d(x, y)$ so we may take $\delta = \epsilon$ in the definition of continuity.

7. $\bar{B}(a, r) \subset T$. The inclusion here may be strict: e.g. if
$$X = \{x \in \mathbf{R}^2 : \|x\| \leq 1 \text{ or } \|x\| \geq 2\} \quad \text{and} \quad r = 2.$$

8. Each individual point is open.

9. Use the inclusion (2) and adapt the proof of Proposition 3.

12. Verify that the mapping $g : \mathbf{R} \to \mathbf{R}^2$ defined by $g(x) = (x, f(x))$ is continuous and then use Theorem 2. Cf. the proof of Theorem 5.

14. Suppose that there are no limit points. To each point x of X there is an open set G_x containing x but containing no other point of X. As X is compact, a finite selection of these G_x's cover X — a manifest contradiction.

16. f is a quadratic polynomial in the 'variables' $x_1, \ldots, x_k$ and hence is continuous.

Let $B = A - \lambda I$. Then $xBx^t = xAx^t - \lambda xx^t \leq 0$ for all x on the unit sphere and hence for all x in $\mathbf{R}^k$ by homogeneity; moreover $cBc^t = 0$. For any real number r and any x in $\mathbf{R}^k$, $(c + rx)B(c + rx)^t = 2rcBx^t + r^2 xBx^t \leq 0$. This implies that $cBx^t = 0$ for all x in $\mathbf{R}^k$ and hence that $cB = 0$ (taking $x = (1, 0 \ldots), (0, 1, 0, \ldots), \ldots$ in turn).

17. Suppose on the contrary that $Y = C \cup D$ where C, D are non-empty disjoint open subsets of Y. We may find c in C, d in D and find a path f in Y whose initial point is c and whose terminal point is d. The image $f([a, b])$ of the path is connected (Theorem 2) but is 'disconnected' by the open sets C, D. This contradiction shows that Y must be connected.

20. $\mathcal{T}_1 = \{\varnothing, \{0, 1\}\}, \quad \mathcal{T}_2 = \{\varnothing, \{0\}, \{1\}, \{0, 1\}\}, \quad \mathcal{T}_3 = \{\varnothing, \{0\}, \{0, 1\}\},$
$$\mathcal{T}_4 = \{\varnothing, \{1\}, \{0, 1\}\}.$$
Of these, only $\mathcal{T}_2$ (the discrete topology) satisfies O 4.

23. (i) Let C be a closed subset of X. Augment an open covering of C by $X \setminus C$ to cover X and argue as in part (ii) of the proof of Theorem 6.

(ii) Let B be a compact subset of Y and let $a \in Y \backslash B$. To each point b of B there exist disjoint open sets G_b, H_b such that $b \in G_b$, $a \in H_b$. As B is compact, a finite selection $G_{b_1}, ..., G_{b_n}$, say, of the G's cover B. Then the open set
$$H_{b_1} \cap H_{b_2} \cap ... \cap H_{b_n}$$
contains a and has no points in common with B. This shows that $Y \backslash B$ is open and hence B is closed.

(iii) By (i), (ii) and Theorem 8, f carries any closed subset of X onto a closed subset of Y. Hence f carries any open subset of X onto an open subset of Y; in view of Proposition 3 (or the *definition* of continuity in the general case), f^{-1} is continuous.

REFERENCES

[1] Apostol, T. M. *Mathematical Analysis*. Addison Wesley, 1957.
[2] Asplund, E. and Bungart, L. *A First Course in Integration*. Holt, Reinhart and Winston, 1966.
[3] Baire, R. Sur les fonctions des variables réelles. *Ann. Mat. Pura Appl.* (3) **3** (1899), 1–122.
[4] Burkill, J. C. *The Lebesgue Integral*. Cambridge University Press, 1961.
[5] Clarkson, J. A. Uniformly convex spaces. *Trans. Amer. Math. Soc.* **40** (1936), 396–414.
[6] du Bois-Reymond, P. Untersuchungen über die Convergenz und Divergenz der Fourierschen Darstellungsformeln. *Abh. Akad. München*, **12** (1876), 1–103.
[7] Fejér, L. Untersuchungen über Fouriersche Reihen. *Math. Ann.* **58** (1904), 51–69.
[8] Goffman, C. *Real Functions*, Volume 8 of the Complementary Series in Mathematics. Prindle, Weber and Schmidt, 1953.
[9] Halmos, P. *Introduction to Hilbert Space*. New York, Chelsea, 1951.
[10] Hausdorff, F. *Grundzüge der Mengenlehre*. Leipzig: Veit, 1914.
[11] Lebesgue, H. Intégrale, longueur, aire. *Ann. Mat. Pura Appl.* (3) **7** (1902), 231–59.
[12] Lebesgue, H. *Leçons sur l'Intégration et la Recherche des Fonctions Primitives*. Paris: Gauthier-Villars, 1904.
[13] Lebesgue, H. Recherche sur la convergence des séries de Fourier. *Math. Ann.* **61** (1905), 251–80.
[14] Lebesgue, H. *Leçons sur les Séries Trigonometriques*, Paris, 1906.
[15] Levi, B. Sopra l'integrazione delle serie. *Rend. Ist. Lombardo Sci. Lettere*, (2) **39** (1906) 775–80.
[16] Loomis, L. H. and Sternberg, S. *Advanced Calculus*. Addison Wesley, 1968.
[17] Mendelson, B. *Introduction to Topology*, Blackie, 1963.
[18] Newman, M. H. A. *Elements of the Topology of Plane Sets of Points*. Cambridge University Press, 1961.
[19] Reid, C. *A Long Way from Euclid*. Routledge and Kegan Paul, London, 1965.
[20] Riesz, F. Sur les systèmes orthogonaux des fonctions. *Compt. Rend. Acad. Sci. Paris*, **144** (1907), 615–19, 734–6.
[21] Riesz, F. and Sz.-Nagy, B. *Functional Analysis*, translated by L. F. Boron. New York: Frederick Ungar, 1955.
[22] Rogosinski, W. W. *Volume and Integral*. Oliver and Boyd, 1962.

REFERENCES

[23] Royden, H. L. *Real Analysis*. New York: Macmillan, 1963.
[24] Rudin, W. *Real and Complex Analysis*. McGraw-Hill, 1966.
[25] Solovay, R. The measure problem. *Amer. Math. Soc. Notices*, **12** (1965), 217.
[26] von Neumann, J. *Mathematical Foundations of Quantum Mechanics*, translated by R. T. Bayer. Princeton University Press, 1955.
[27] White, A. J. *Real Analysis: an introduction*. Addison-Wesley, 1968.
[28] Whittaker, E. T. and Watson, G. N. *A Course of Modern Analysis*, Fourth Edition, Cambridge University Press, 1935.
[29] Zygmund, A. *Trigonometric Series*, Second Edition, Cambridge University Press, 1968.

INDEX

Absolutely continuous function, 67
Absolutely convergent series, 5
Affine mapping, 139–44
Almost everywhere, 32, 77
  convergence, 168
Angle, 174
Archimedes, Axiom of, 3n, 14, 164
Area, 71, 82, 87

Baire classes, 122
Ball
  open, 82, 224
  closed, 82
Banach space, 221
Bessel's equation, 198
Bessel's inequality, 198
Bolzano–Weierstrass Theorem, 239
Bonnet's Mean Value Theorem, 134, 210
Borel sets, 145
Boundary, 227
Bounded Convergence Theorem, 110, 112
Bounded set, 233
Bounded variation, 65, 209
Brouwer's Theorem, 158

Cantor, 16, 17
  ternary set, 20, 69, 145
Cauchy
  General Principle of Convergence, 116, 163, 169
  sequence, 164
Cesàro, 211, 218
Characteristic function, 23, 72
Clarkson's inequalities, 181
Closed
  ball, 82
  disk, 82
  set, 135, 227
Closure, 193, 227
  of an interval, 71
Completeness
  of R, 2–13, 162–4
  of L^1, 42
  of L^p, 170
  of a metric space, 164
  of a set in L^2, 196
Compact set, 135, 234–6
Complex numbers, 13

Connected set, 229–33
Continuous function, 46–50, 79–81, 228–38
Continuously differentiable function, 156
Convergence
  of a sequence, 2–10
  of a series, 5
  pointwise, 168–71, 202–14
  almost everywhere, 168–71
  in mean (strong), 168–71, 207
Convex set, 176
Countable set, 15
Cross-section, 85, 87
Cube, 71

Darboux's Theorem, 52
Decimal expansion, 5–7, 10, 17
Decreasing
  sequence, 2
  function, 48
Dedekind's Theorem, 14
Dense, 15, 196
Density function, 146–61, 185–6
Derivative, 55–69
  measure, 152–4
  linear, 155–9
  directional, 155
  partial, 155
Determinant, 144
Disk
  open, 82, 89
  closed, 82
Differentiation under the integral sign, 118
Dimension, 192
Dirichlet kernels, 205–13, 216–17
Distance, 1, 70, 155, 223
Dominated Convergence Theorem, 106, 109, 113
Dual spaces, 166, 184, 187
du Bois-Reymond, 214

Eigenvalue, maximum, 240
Elementary figure, 75
Erlanger Programm, 134
Euclidean space, 70–92, 124 *et sqq.*, 219 *et sqq.*
Exchange Lemma, 193
Exterior, 227

Fatou's Lemma, 110, 116
Fejér, 211
  kernels, 212–14
  Theorem, 213, 214
Finite dimensional subspace, 191
Fourier series, 188, 194–9, 202–19
Fubini's Theorem, 83, 90, 123, 142
Function, 22
  absolutely continuous, 67
  of bounded variation, 65, 209
  continuous, 46–50, 79–81, 228–38
  continuously differentiable, 156
  generalised step, 128
  increasing, decreasing, monotone, 48
  measurable, 120–4, 128–33
  simple, 128, 130
  step, 25–30, 72–6.
Fundamental Theorem of the Calculus, 56, 67, 112

Gamma function, 111
Geometry, 134
  of L^2, 172–81
Gibbs' constant, 218
Gibbs' Phenomenon, 207, 218
Gram–Schmidt Theorem, 191

Hausdorff space, 238
Heine–Borel Theorem, 34, 233
Heisenberg matrix theory, 200
Hermite functions, 201
Hilbert space, 221–2
Hölder's Inequality, 167
Homeomorphism, 144, 238
Hyperplane, 71

Increasing
  sequence, 2
  function, 48
Infimum, 9
Inner product, 172
Integers, 1
Integrable
  function, 37, 79
  set, 125
Integral
  Lebesgue, 37–43, 79–92
  definite, 44–50
  indefinite, 54–62
Integration
  by parts, 58, 103
  by substitution, 59, 104
Intermediate Value Theorem, 231
Interior, 227
Interval, 1, 2, 70, 71
  length, 2
  measure, 71

Inverse Mapping Theorem, 159
Isometry, 137–44

Jacobian
  matrix, 155
  formula, 158
Jordan's Theorem, 209

Klein, Felix, 134

L^{inc}, 33–6, 41, 46, 78
$L^{\text{inc}}(A)$, 126
L^1, 37–43, 47–50, 78–81 et sqq.
$L^1(I)$, 40, 79
$L^1(A)$, 126
L^p, 162–222
$\mathscr{L}^p$, 165–6
Laguerre polynomials, 200
Lebesgue, ix et sqq., 202
  integral on $\mathbf{R}$, 22–69
  integral on $\mathbf{R}^k$, 70–92
  measure, 124 et sqq.
Legendre polynomials, 200, 202
Length, 2, 24, 28, 87
Levi, Beppo, 93, 107
Limit point, 226
Linear
  space, 22, 72, 165, 199, 220
  operator, 72
  mapping, 138
  subspace, 176, 220
  functionals, 182–8
  dependence, independence, 192, 194
Lipschitz Condition, 69
Localisation Principle, 206
Lower bound, 3, 9

Matrix, 139
  unit, invertible, elementary, diagonal, 140
Mean, arithmetic, geometric, 62
Mean Value Theorem, 237
  for integrals, 41
Measurable
  function, 120–4
  set, 124
Measure, 75, 87, 124–46
  of a bounded interval, 71
  derivative, 152
Metric, 223
  space, 163, 220, 223
  complete, 164
  discrete, 224
Minkowski's Inequality, 167, 172
Monotone Convergence Theorem, 93, 96, 103, 106

INDEX

Monotone
  sequence, 2
  function, 48

Neighbourhood, 239
Nested intervals, 14
Non-measurable set, 131
Norm, 70, 154, 165, 199, 220
Null set, 18, 77

Open
  ball, 82, 224
  set, 135, 225, 237
  relatively, 230
Ordered field, 12
Ordinate set, 88, 133
Origin, 1
Orthogonal, 174
  complement, 180
  projection, 180
Orthonormal set, 191–201

Parallelogram, 175
Parseval's equation, 199–200
Pointwise convergence, 168–71, 202–14
Polar transformation, 146, 159
  cylindrical, 159
  spherical, 160
Positive, 1
Projection, 179
Pythagoras' Theorem, 6, 7, 174

Radon–Nikodym Theorem, 152, 185
Rational numbers, 1
Real line, 1
Relatively open set, 230
Riemann integral, 47, 51–4, 83
  improper, 106
Riemann–Lebesgue Lemma, 115, 204
Riesz's Theorem for L^2, 183
Riesz–Fischer Theorem, 199
Rolle's Theorem, 236

Scalar product, 137, 199, 221
  (inner product), 172
Schrödinger, 200

Schwarz' Inequality, 137, 173
Segment, 174
Sequence, 2
  increasing, decreasing, monotone, 2
  upper, lower, 107
  pointwise convergence, 168–71, 202–14
  convergence almost everywhere, 168–71
  convergence in mean, 168–71, 207
Series, 4
  convergence, 5
  absolute convergence, 5
  convergence in L^p, 169
Sets, union, intersection, difference,
  symmetric difference of, 23
Simple function, 128–30
Step function, 25–30, 72–6
  generalised, 128
Strong convergence, 168–71, 207
Summable $(C, 1)$, 211
Supremum, 9, 12
Symmetric difference, 23

Tonelli's Theorem, 123
Topological mapping, 144, 238
Topology, 223–40
  general, 226, 237
Triangle inequality, 138, 165, 173, 223
Trigonometric polynomial, 190

Uniform continuity, convergence, 218
Upper bound, 3, 9, 12

Variation
  bounded, 65, 68
  total, 68
  positive, negative, 68
Vector space, 220
Volume, 71, 73, 83, 87

Weierstrass' Approximation Theorem, 218
Well-ordering of positive integers, 8

Zorn's Lemma, 221

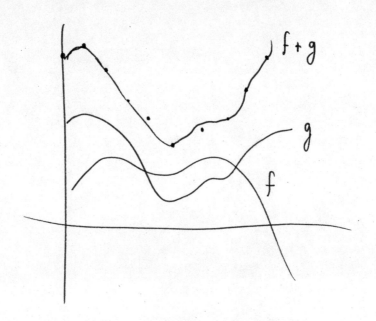